WHAT IS SCIENCE?

An Introduction to the Structure and Methodology of Science

V. James Mannoia

Westmont College
Santa Barbara

UNIVERSITY PRESS OF AMERICA

LANHAM • NEW YORK • LONDON

University Press of America,™ Inc.

4720 Boston Way
Lanham, MD 20706

3 Henrietta Street
London WC2E 8LU England

ISBN (Perfect): 0-8191-0989-4
LCN: 79-47988

Additional Information:

Typeset in 10 point Helvetica by
Murray Typesetting and Editorial Service
Santa Barbara, California

Illustrations by
Sally Hartman

All University Press of America books are produced on acid-free
paper which exceeds the minimum standards set by the National
Historical Publications and Records Commission.

To Ellen

ACKNOWLEDGMENTS

I want to thank a number of people and groups who helped make this book possible. The material has been endured by my students; first in the Science Keystone course at Grove City College in 1975, and subsequently by students in my Philosophy of Science courses. Richard Blackwell, a much respected mentor, encouraged me to expand the material and to publish it. I am grateful to the National Endowment for the Humanities for funds and opportunity to work without interruption on the final revision during the hot summer of 1978 at Johns Hopkins. I must thank Sandra Hubert for her comments on the history of science chapter. Dave Downing not only read the entire typescript on rather short notice, but gently coaxed me towards intelligibility of style. (I fear his efforts have not been entirely successful, but that is not his fault.) Louise Phillips in Baltimore and Betty Bouslough in Santa Barbara worked very hard in typing drafts, footnotes, and bibliography. My thanks go to Tom Andrews, our academic dean, for uncovering (or inventing) some obscure fund to aid in the final preparation costs. Kristen Cantrell did a very efficient job on the index and Sally Hartman's illustrations are worth thousands of words. John and Ginny Murray did the final preparation and not only gave it that priceless personal touch but tolerated changes with the patience of Job. Finally, I thank my wife, Ellen, who not only did "dirty" details and supported me financially during the first summer of writing, but knew I could do it all along.

CONTENTS

INTRODUCTION

The following material is intended to help bridge the wide gulf between what C. P. Snow has termed "two cultures."[1] On one side of the "Snow Gap" is what may be loosely termed 'the scientific community,' including scientists, engineers, technicians and so on. On the other side of the gap is the humanities or liberal arts community consisting of those involved in philosophy, literature, fine arts, and even certain non-"natural" sciences. This gap is apparent on university campuses between students in science-related disciplines and those in the liberal arts. Without repeating Snow's own work, let it suffice here to say that the "gap" in question involves a major inability to communicate. This is produced in part by the differing "languages," purposes, and methods of each group. It is further widened and reinforced by two attitudes against which this essay is more specifically addressed.

First, it is quite common for members of the liberal arts community to stand in fear or fascination of the scientific community. This attitude is the product of our age—an age when the tangible fruit of science is everywhere. There are exceptions, and perhaps these exceptions are becoming more numerous today[2] but it remains true that the very specialized technical languages of the sciences can easily intimidate the uninitiated. It is not uncommon to hear a liberal arts student say, "Oh you're a *physics* major, you must be a brain!" Or perhaps the student complains, feeling somehow inferior, that science is "too complicated for me." This attitude needs to be overcome. Non-scientists must not fear or be intimidated by science. They should learn instead to appreciate it and even study it in order to overcome a fear born of ignorance and to equip themselves as citizens in a scientific age.

The second attitude perpetuating the "Snow Gap" is the unjustified sense of superiority too often exhibited by members of the scientific community. It may stem from the uncritical pride scientists take in their success of the past hundred years. The attitude is encouraged by an educational system teaching that science alone deals in facts; that science alone is objective. Nothing stands in the way of

1

communication more effectively than an attitude of superiority. So the student of science may be heard to say, "Why should I listen to artists and poets? They know nothing about the real world, the one that really matters. They can't even agree among themselves, what could they say to me? If it cannot be measured, it is too subjective and not useful at all." Again there are exceptions, especially today as complex problems arise in society, born of technology itself. There is a growing realization that science often creates as many new problems as it solves old ones. But the superior attitude remains and must be overcome. Every scientist must study the assumptions, goals, and methods of science carefully enough to see its limitations and inherent uncertainty. Science should not become a sacred cow.

This book is directed to the problem of eliminating these two attitudes. The basic approach is to provide a better understanding of the workings of science to students on both sides of the gap. Ignorance can only widen the gap by cultivating the mistaken attitudes of fear and superiority. What follows then amounts to a brief introduction to the philosophy of science. It is aimed at both those in sciences and those in the humanities and presupposes no acquaintance with philosophy. It has been used as the text for one part of an interdisciplinary course in science but could also presumably be used for part of a philosophy course or, with suitable supplementary readings, as the basic text for an undergraduate course in Philosophy of Science.

CHAPTER ONE

The Formation of Scientific Ideas: Scientific Method

INTRODUCTION

Philosophy of Science

The Philosophy of Science is, generally speaking, an attempt to step back from science and look at it as a whole. Without this perspective "it is easy to miss the forest for the trees." Non—scientists are often not interested at all in science and scientists are usually too close to it, too involved with specific "trees," to notice what science is all about. The philosophy of science is then an important tool for providing scientist and non-scientist alike with a perspective of science which minimizes mutual misunderstanding.

When we step back to look at science we discover that it is both something you *do*—as scientists work at doing science—and something you *have*—as the laws of physics which can be found in a textbook. Science is both an *activity* and a *collection of facts*. Put in another way, 'science' can be thought of as though it were a verb, or on the other hand as a noun. This important two-fold distinction regarding science I shall call the 'dynamic' and 'static' perspectives of science.

Main Points: Science is Common Sense, Not Magic

This distinction of perspectives provides a place to begin in the study of science. This chapter and the next concern the study of science from the dynamic perspective. Chapters Three and Four are investigations of science from the static perspective. Finally, Chapter Five contains a discussion of the limitations of science both in its method (dynamic) and its result (static).

To begin with, it is useful to ask in general how one would go about studying any activity.[1] If, for example, the activity is a stageplay, one might *study* it (as opposed to just enjoying it as performed) by examining the script. If the activity is a television show, producers often use a "story board" diagramming the plot and various scenes on paper. If the activity is a computer program designed to act on certain data and produce information, we study the computer's activity with a flow chart. The common element in any study of an activity is the *time* dimension. Without taking time into account the study loses the essence of the activity it studies. A static "blueprint" of a machine's inner structure tells us little about how that machine works, until we add some idea of which parts move first, what way, and how fast.

Science, like any activity, has a peculiar character or method. Our concern here then is to investigate this method of science; the way scientists act to *form* scientific ideas. We are not concerned with the content of those ideas, or even—until Chapter Four—with their use.

There are two *main points* to be made in what follows. 1. The activity of doing science is very much like everyday problem solving. To this extent, the scientific method is just common sense and we are all scientists.[2] 2. The activity of forming scientific ideas is not some kind of mechanical process that proceeds automatically if you only know the secret rules. Rather, it requires *both* careful attention to precise rules *and* a vast dimension of creativity. In making these main points, I hope to take some of the mystery out of what scientists do and help them recognize their limitations—for their good as well as our own.

EXAMPLES OF SCIENTIFIC ACTIVITY

Let us examine some important examples of the scientific activity, in order to provide a context in which to illustrate some of its important features.

The first illustration comes from the history of medicine. It is a description of the discovery of the role of micro-organisms in infection. Ignaz Semmelweiss was born in Budapest in 1818. After receiving his medical degree in Vienna he was appointed assistant at the obstetric clinic there. He was surprised to discover that an abnormally large number of the women who delivered their babies in his own First Maternity Division developed a serious and often fatal disease known as "childbed fever." As many as 11% of the women admitted actually died of the disease. The problem was especially troublesome to Semmelweiss because patients in the adjacent Second Maternity Division suffered far fewer fatalities; in fact fewer than 3% died of the disease there.

Semmelweiss' attempt to explain this problem led him to consider a variety of hypotheses. Some were quite inadmissable because they ignored well-known facts but others he examined more carefully. The process by which he arrived at a solution is a case in point for the formation of scientific ideas.

Among the explanations he considered was the popular view that "epidemic influences" or "atmospheric-cosmic-telluric-changes" enveloped whole regions causing the death of pregnant women. Semmelweiss rejected such theories on the evidence that had it been true then: 1. surely the entire city would have suffered, yet the fever occurred only rarely in Vienna or its environs while it raged in the hospital; 2. surely an "epidemic" even if selective enough to single out the hospital would strike both First and Second Division wards equally hard, yet the Second division was spared.

Another theory held that overcrowding caused the deaths. But this too was rejected since on that view the *Second* Division mortalities would be much higher. Women desperately sought assignment to the Second Division out of fear of the First Division's high mortality rate, so the Second Division was much more crowded than the First.

Two other hypotheses were thrown out when Semmelweiss noted no differences in diet or general care of patients between the two divisions.

In desperation, Semmelweiss even considered psychological explanations. Some people noted that whenever a patient was dying a priest was called to administer last rites in a special sick room. In the Second Division the priest had direct outside access to the sick room there, but in the First Division the priest passed through five wards of women to enter this room. It was held that the grim procession with funeral chant and ringing bell terrified the women causing the fever. If true, Semmelweiss reasoned, eliminating the procession would reduce the deaths. Yet, when he persuaded the priest to eliminate the bell and come a roundabout way silently and unobserved, the First Division death rate did not decrease.

Still groping, Semmelweiss observed that in the First Division the women were delivered lying on their backs; but were delivered on their sides in the Second Division. However, changing the First Division procedure to lateral delivery brought no change in fever deaths.

In early 1847, an accident gave Dr. Semmelweiss a decisive clue. Another physician on the ward, Dr. Kolletschka, was performing an autopsy with the aid of a young medical student. During the autopsy, Kolletschka was cut by the inexperienced scalpel of the student. Kolletschka died after an agonized illness during which he showed many of the symptoms of the childbed fever. The similarity of symptoms suddenly suggested a connection of autopsies and childbed fever. Although there was as yet no concept of the role of germs in infection, Semmelweiss realized Kolletschka had died because "cadaveric matter" or "death tissue" had entered and "poisoned" his blood. The similarity of Kolletschka's death and childbed fever suggested by analogy that perhaps the women also died from poisoning by "cadaveric matter." If true, this theory would imply that the women must somehow be having contact with such matter; i.e., what we today would call "infection." This prediction was confirmed when Semmelweiss realized that he and the medical students carried the matter to the women on their hands. They would come to the maternity ward directly from the dissections in the autopsy room to examine women in labor after only washing superficially. Who would have thought hand washing was relevant to childbed fever?

Semmelweiss tested his idea further reasoning that, if true, the theory implied that by chemically destroying the cadaveric matter in thorough washing, the deaths should decrease. He ordered a new hand washing procedure as an experiment. All medical students were instructed to wash with chlorinated lime before any examination. The fever deaths promptly declined and in the following year were below even those of the Second Division.

The theory also accounted for the Second Division's originally lower mortalities. The patients there were examined by midwives whose training did not include anatomy by dissection of cadavers! The theory also explained why women who arrived with babies in arms—having given birth prematurely in the street enroute—were far less likely to contract the fever. They were less likely to be examined once in the hospital.

The work of Semmelweiss was among the most important contributions to modern concepts of hygiene, sterilization, infection and microbial transmission of disease.[3]

A second important illustration of the scientist at work is Isaac Newton. Born on Christmas Day in 1642, Newton was perhaps the greatest genius of a century of genius. He shared the stage of progress in the 17th century with such men as Kepler, Galileo, Leibniz, Descartes and many others. Perhaps best known for his theories of universal gravitation and motion, which gave a systematic and revolutionary explanation of the physical universe, Newton began his inquiries with an interest in light.

Since the time of Aristotle, it had been common knowledge that a prism produces colored light. However, in 1666, the prevailing explanations of this phenomenon were quite inadequate. The Phenomenon of Colors was usually explained by the Alteration Theory. In this view, white light passing through the thin edge of a prism is altered (darkened) a little so it only becomes red. It is altered a little more where the glass is thicker and becomes green. It is altered still more where the prism is thickest and becomes blue. To Newton this theory *explained* absolutely nothing, however plausible it may have sounded. He had observed a very obvious thing which the theory couldn't begin to explain: when sunlight is permitted to fall on a prism through a small hole, the resulting spectrum is elongated; yet the sunlight comes in as a circular disc. What for generations was unnoticed or unimportant was a problem for Newton. It takes a powerful mind to see the need to explain the obvious. Newton himself described the act of discovery whereby he explained the problem of the elongated spectrum. That description is perhaps more significant for what it omits than for what it includes. This will become more clear in our discussion of Discovery below.

> I procured me a Triangular glass prisme, to try therewith the celebrated Phaenomena of Colors. And in order thereto having darkened my chamber, and made a small hole in my window shuts, to let in a convenient quantity of the sun's light, I placed my Prisme at his entrance, that it might be thereby refracted to the opposite wall. It was at first a very pleasing divertisement, to view the vivid and intense colors produced thereby; but after a while applying myself to consider them more circumspectly, I became surprised to see them in an *oblong* form: which according to the received laws of Refraction, I expected should have been *circular.*

> I saw . . . that the light, tending to [one] end of the image [spectrum], did suffer a refraction considerably greater then the light tending to the other. And so the true cause of the length of that image was detected to be no other, then that *Light* consists of *rays differently refrangible,* which, . . . were, according to their degrees of refrangibility, transmitted towards diverse parts of the wall.[4]

This discovery was the theory of the composite character of white light. Newton had seen that if white light was understood to consist of many different colored rays, each of which could be bent (refrangible) at different angles by the prism, one could then not only account for the prism's spectrum (as could the alteration theory) but for the spectrum's elongation as well.

Newton tested his theory in at least two ways. First his theory implied that once separated into its respective colors, light would not be changed by further refraction. The alteration theory would imply on the other hand that a second prism should further alter the light and produce new colors. Red should become green or blue and so on. But after trying the experiment Newton noted:

> When any one sort [color] of rays hath been parted from those of other kinds, it hath afterwards obstinately retained its color, notwithstanding my utmost efforts to change it.[5]

Second, Newton saw that if true, his theory would imply that the separated colors of a prism spectrum when recombined would once again produce white light. This too was confirmed by Newton's experiments. Reporting this result, Newton said,

> I have often with admiration beheld, that all the colours of the prisme being made to converge, and thereby to be again, mixed, reproduced light, intirely and perfectly white.[6]

While the composite character of white light is elementary school science today, it began like all theories as a tentative guess at a difficult problem.

GENERAL CONSIDERATIONS: SCIENCE IS A GAME

From the illustrations above, one general feature of the scientific activity emerges. It is an interplay between boundaries of observation and theory. First, scientists are clearly concerned with what can be observed. Semmelweiss could not ignore the fact of higher mortality rates in his division than in the other. He could not pass off as irrelevant the *observation* that a priest could be seen and heard passing through the maternity wards. Science deals in what can be experienced with the five senses. It is an empirical and experimental activity. This principle is the very strength of science. No self-respecting scientist would deny its centrality. He or she would never assert that, "My mind is made up. Don't confuse me with the facts!" It is at the heart of science to be concerned with observations made of the world. As we shall see the scientific activity both begins and ends there.

Second, however, scientists are equally concerned to form ideas about these observations. Whether we call them theories, or hypotheses, or just guessed answers, these ideas show that scientists are also concerned with *concepts*. It is not enough for a scientist to notice things with the senses. Unless there are ideas formed about these observations, the activity of science is no more than just "experiencing."

These two levels of concern, observational and theoretical, form the boundaries or "playing field" for the scientific enterprise. But they differ not only because the former is concrete and the latter abstract. The Observational/Empirical level is also usually the level of particulars while the Theoretical/Conceptual level is usually general. By this I mean that the distinction of these two levels of concern is more than just the difference between *seeing* a dog and *thinking* about a dog or between *feeling* a ball and *imagining* a ball. It is also the difference between seeing a *particular* dog Rover and thinking about what all dogs *generally* have in common. Put simply, the point is that science is an activity concerned with both particular observations and with general concepts.

10

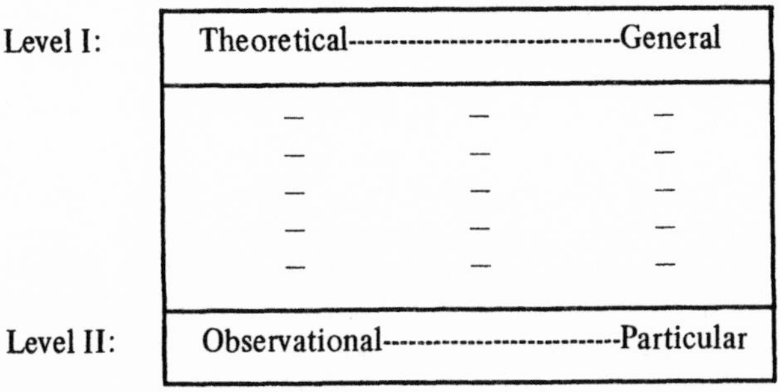

Level I:	Theoretical-----------------------------General
	— — —
	— — —
	— — —
	— — —
	— — —
Level II:	Observational--------------------------Particular

Level I is also called Conceptual, Hypothetical, Abstract
Level II is also called Empirical, Experimental, Concrete

Figure 1: **The Playing Field of Science.**

Having discussed the "playing field" of science, we can now more easily describe the object of the "game" itself. The purpose of science as an activity is to *form conceptual generalizations about the many particulars of empirical experience.*[7] There are of course innumerable definitions of science. One might also say for example that it is the business of science to form theories about observations. Einstein came quite close to our definition when he said, "Science is an attempt to make the chaotic diversity of our sense experience correspond to a logically uniform system of thought."[8] Each of these definitions clearly shows how science is involved on both levels of concern. With this in mind let us turn directly to the question of how science accomplishes its task. Specifically, how do the methods of the scientific activity tie together its two levels of concern in accordance with its purpose?

SCIENTIFIC METHOD

Common Sense
One main point of this chapter is that the activity of science is much like the processes all men use every day in solving problems. To this extent we are all scientists.[9] Whether we think of it as "common sense" or something else, and whether we are conscious of it or not, each of us, generally approaches problems in much the same way.

To define "common sense" is probably impossible. But there are certain features of our problem solving approach that are quite apparent. The thought process usually begins with a *problem.* That problem can be as simple as the question, "Where is my roommate?" or as complex as solving a differential equation. In any case the problem is

11

some particular item of our experience that requires explanation. Frequently the problem is a problem because it is anomalous; i.e., does not fit our expectations.

When presented with such a problem our mind almost immediately begins to imagine *possible answers* to our problem and to wonder if they are true. These answers take the form of ideas (theories) in our minds. For example, you might wonder almost without thinking whether your roommate is in the room next door or at class. We reject, though usually unconsciously, any answers that are incompatible with our existing expectations about the world. For example, we reject as ridiculous (for most roommates), probably before it ever consciously surfaces, the possibility that our roommate is in jail or on the moon. (Regardless of any *wishes* to the contrary!)

But our wondering whether a given acceptable possibility is true leads us next to *test* that theory. We quite reasonably deduce that *if* our roommate is in the room next door *then* we should see him or her if we too enter that room with our eyes open. We may or may not choose to perform this little experiment depending on how badly we want to know if our theory is true. In any case this testing of our theory is one which appeals to the particular observations of our sense experience as the ultimate criteria for judging truth. Like science, common sense is emphatically empirical.

Finally, in the event we do enter the room next door the *result* is that we will either see our roommate or not. In the first case we consider our problem solved. In the second case we must go back to imagining another answer, the "in class" guess or "at lunch" possibility. The process repeats itself.

Naturally, we are usually unaware of going through such a series of distinct steps, (Problem, Possible Answer, Test, Result). In fact the process is often considerably more complex than this simple description suggests. Nevertheless the procedure is familiar and this brief sketch is sufficient for our purposes.

Analysis Into Three Processes

A more careful examination of the scientific activity—and in fact of any *activity*—focuses not on the *steps* followed but on the thought *processes* involved. Such an examination reveals three important processes which will serve as the framework for our discussion in the balance of this chapter. These are the processes of *Discovery, Prediction, and Confirmation.*[10]

1. Discovery
    a)   *Nature* – When faced with the anomalous mortality rate in his maternity division, Semmelweiss immediately sought

an answer to his problem. What he sought may be called an answer, a solution, an explanation, an hypothesis, or even a theory. In any case, the process of discovery is the activity of moving from problem to a possible answer. The key word here is *possible,* because discovery only produces working hypotheses that may or may not be true. But one must begin somewhere—better to start off wrong than never to start at all for fear of being mistaken. Semmelweiss "discovered" many possible answers to his problem; the theory that delivery "position" caused the deaths, the theory involving the entrance of a priest, the theory of "cadaveric matter" and so on. Each must be considered a discovery— however far-fetched—because it *did* account for his situation.[11]

The process of discovery is an activity moving from the level of *particular observations* to *general theories.* In other words, Semmelweiss was not searching for an explanation for only one death, he was searching for a general explanation of many particular deaths. This general character of the theories sought may be one way of distinguishing the scientific process from the common sense procedure outlined above. The general character of the theories sought is evident from the purpose already stated for science as an activity: "To form conceptual *generalizations* about the many particulars of experience." The nature of discovery may be conveniently diagrammed on the two-level "playing field" of science.

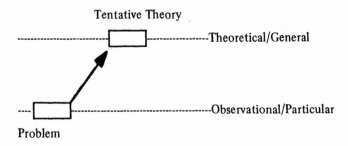

Figure 2: Discovery

b) *Logic: Abductive* — Because it is a process of human thought, discovery may be said to have a "logic" or structure.[12] The name given to this logic is 'abduction.'[13] The process is unrelated to kidnaping and, as you might gather from the examples of Semmelweiss, is more akin to educated *guessing* than nearly anything else.

c) *Misconceptions* — Before we examine more closely the character of discovery and abduction, we must consider and reject two common misconceptions of discovery that distort the layman's understanding of science and of the men who practice it.

How does abduction take the scientist from problem to tentative theory? Two extremes must be avoided in answering this question. First, discovery is *not* some mechanical process that occurs automatically, according to some fixed set of rules when one has collected enough observations or facts. This view presupposes an exaggerated empiricism. On the other hand, *neither* is discovery a matter of sheer, arbitrary guess work. This view reveals an exaggerated rationalism as explained below.

Few people make the second mistake regarding discovery. Unfortunately, many make the first mistake and conclude by worshipping science as some kind of infallible answer machine. Scientists are thought to be mechanical robots who can't miss if they just follow the logic of discovery. Let's look at each of these mistakes in turn.

i. The Christopher Columbus Fallacy — Persons who make the first mistake conceive of discovery as a matter of uncovering answers that are already lying around waiting to be found. Those who commit this Christopher Columbus Fallacy assume that facts speak for themselves.[14] If we just collect enough facts, the answers to our problems will appear. They see scientists as if they were ants rummaging around to collect and record all possible tidbits of sense experience. It is as if scientists were robot photographers who unlike the human artist adds nothing of his own to his picture. The expected product of science on this view is little more than an encyclopedia of facts.

In the history of science, Francis Bacon is said to have held this mistaken view of discovery.[15] Whether or not he thoroughly misunderstood discovery is a question beyond the scope of this chapter. It is true, however, that he believed that discovery proceeded mechanically by a complex set of fact-gathering and list-making rules. It should be noted that his acclaimed "method" of discovery did not lead to a single important new theory in science.

Perhaps this Fallacy can be best illustrated by a simple story told by the philosopher Karl Popper.

"Popper's Fable"

Once upon a time there lived a man who wished to give his whole life to science. Noble creature that he was, this man sat down with pencil in hand and recorded in a notebook everything he could observe. He included everything, from the weather, the racing results, and the levels of cosmic ray bombardment, to the stock market reports and the appearance of all the planets. He also did not neglect to record the attendance of all students at weekly lectures. Our dedicated observer continued this job every day for the rest of his life.

He had compiled, by the time of his death, the most comprehensive record of nature ever made in the history of mankind. When he died, he was so certain that his life had been well spent for the cause of science, he donated his notebooks (there were hundreds of them as you might suppose) to the American Academy of Science.

But did the Academy bother to thank him for his donation and efforts? Did the Academy even accept his donation? No! They refused even to open his notebooks, because they knew without looking that they would contain only a jumble of disorderly and useless items.[16]

The point of course is that science demands more than just observation. In order to decide which observations are worth making and which are irrelevant, the scientist must use an hypothesis (theory) to guide his activities. To paraphrase the 19th century philosopher Immanuel Kant: if theories without observation are empty, then observations without a theory are blind.

The point of Popper's Fable is that "Facts do *not* speak for themselves." To generalize facts into an appropriate theory one must first decide which facts are *relevant* to the problem at hand. But this is no easy job. Semmelweiss could not know which facts were important in his search. Was the priest's presence a fact on which to build a theory or just irrelevant? Was the delivery position relevant? How was he to know that hand washing *was* a crucially relevant fact? This fact did not jump out and speak for itself. Its relevance became apparent only after a tentative conceptual framework (theory) was already at work. Discovery then is more than the automatic result of simply collecting facts or observations according to some mechanical system.

ii.   The Alice-In-Wonderland Fallacy — The second major misconception is the view that discovery is sheer guesswork, arbitrarily generating theories that, like the ideas of Alice-in-Wonderland, are utterly out of touch with the reality of facts.[17] Scientists are sometimes thought to be men, who when faced with a problem, retire to ivory towers where, like spiders, they spin theories like so much silk from the recesses of pure imagination. The resultant theories are merely projections of the scientists' minds, with little relevance to the particular facts associated with the problem.[18] Because of this irrelevance, such theories either cannot or need not be tested against the standard of sense experience. Too much personal contribution to the process (i.e., subjectivity) by the scientist renders the theories useless for the purposes of science.

In summary, the Christopher Columbus Fallacy isolates the scientist at the observational level of concern where he is unable to rise to the necessary theoretical generalizations. On the other hand, the Alice-in-Wonderland Fallacy isolates the scientist at the theoretical level, unable to establish relationship to the observational level of sense experience.[19] Neither conception accurately characterizes the process of discovery.

   d) *Characteristic: Creative* – A more accurate picture of the characteristics of discovery (abduction) is that of balance. Abduction involves both careful attention to observed facts and a crucial element of intuitive imagination. This balance we will call 'creativity' and is the essence of discovery. The creative scientist must show careful appreciation for the problem and have as broad an observational awareness as possible. But then the scientist must be willing to risk error, by making some good imaginative guesses.

Continuing the metaphor of the ants and spiders, one might say that each act of discovery requires the contribution of both the ant and spider in balanced cooperation. Bacon himself, who first used this metaphor to widely *separate* the empiricist and rationalist approaches to knowledge, seemed to have a premonition that the separation must ultimately collapse into an indistinguishable blend of both. He found the activities of the bees more promising as a model for discovery. In

his words:

> But the bee takes a middle course; it gathers its material from the flowers of the garden and of the field, but transforms and digests it by a power of its own.[20]

Abduction is clearly difficult to define except by saying what it is not. You will recall from the stories of Semmelweiss and especially of Newton that little is said of exactly *how* they obtained the tentative theories they tried. Hence what Newton *did not* tell us was significant because he probably *could not* tell us. Abduction is neither mechanical nor arbitrary but a creative balance. It depends upon tacit intuitions born of experience, hunches, accidents, and, perhaps more than any other single factor, upon the use of analogy.

An analogy is a relationship of similarity between two entities or situations. A common role of analogy in discovery is that often a strong similarity exists between *parts* of a problem context and some other area of experience. This other area can then serve as a model from which the scientist draws further characteristics which he is hopeful will *also* apply to the problem context. The assumption of course is that if there are similarities in one area there will be similarities in others. The assumption is not always true but serves the creative scientist as a place to begin in hypothesizing.

Suppose, for example, your problem is to determine the horsepower of the engine in a car placed before you. The car is a white 1966 Ford Mustang convertible. Suppose too that at one time you owned a white 1966 Ford Mustang convertible and that your car had a 289 cubic inch engine. By noting the analogy and using your car as a model, you would probably guess that the answer to your problem is 289 cubic inches. You may be mistaken. The problem car may have a one horse lawnmower engine, a rubberband motor, or no engine at all, but you would be foolish to ignore the analogy and start by guessing something other than 289 cubic inches.

Other examples of analogy at work in discovery are evident in the story of Semmelweiss. It is an interesting exercise to identify the problem, model, and analogies in that example.

e)   *Implications* — It should be remembered first that discovery is a creative process. No discipline holds a monopoly on creativity, and any view which places science at the opposite extreme from fine arts on a scale of "creativity required" is quite mistaken.

Second, if discovery is creative, then scientists must be creative. All too often scientists are viewed as narrow, precise, mechanical sort of people. But real scientists always maintain a balance. They must be precise and mathematical in their attention to detail but imaginative and even daring in their search for solutions. The genius of Isaac Newton underscores this truth. A cursory glance at his *Optics* or *Principia* will testify to Newton's observational and conceptual precision. Yet he was also famous (even notorious) for his temperamental egoism, and imaginative (if unorthodox) work in theology and even alchemy. Good detectives also embody the balance of observational precision and imaginative freedom essential to discovery.

Finally, failure to appreciate this dual or balanced character in good science means unfortunate misconceptions of science among laymen. First the failure makes it possible for science and scientists to be mistakenly worshipped as if they were cold, impersonal, and supremely objective. Second, textbooks of science which neglect this dimension of science, present distorted histories of science and can easily mislead the student by inadequately preparing him for participation in the discipline.

We have understood discovery as a process for obtaining tentative theories or working hypotheses that may very well be wrong. A natural question arises. How does the scientist know whether he is right in his "guess?" The answer lies in the fact that discovery alone is incomplete. It is only a part of the complete activity of science. Let us turn now to the two processes which together serve to test or justify these "educated guesses."[2][1]

2. Prediction

a) *Nature* – Having arrived at a tentative theory or working hypothesis, it remains for the scientist to test its truth. The first step of that test is the process we will call 'prediction.' It consists of moving from theory to test, from hypothesis to experiment. Because most theories are of a general character, prediction is usually a process of moving from *general* to *particular.* In terms of the two levels of concern introduced above, this may be illustrated as follows.

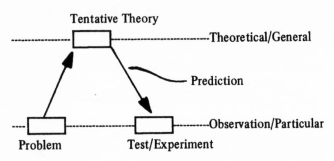

Figure 3: Prediction

18

The purpose of prediction is, of course, to state what consequences must be expected *if* the tentative theory is really true. Predictions are then observable consequences of the theory.

Illustrations of prediction are quite easy to understand. In our example of searching for your roommate, one tentative hypothesis was that your roommate was in the room next door. An observable consequence of that hypothesis is that: *If* you entered that room next door with your eyes open *then* you would see your roommate. Notice how the consequence, a) follows naturally from the theory, b) takes the form of an *observable* consequence, and c) has an *if/then* structure. The observable character and *if/then* structure help us to see how the results of prediction are usually experiments which *can* be performed. In this example the experiment is simply to go next door and look. More frequently, scientific experiments involve lengthy time periods and complex equipment. At times, the experiments may even be mathematical or statistical in nature. Ultimately, however, they are empirical.

Several experiments based on prediction can be seen in the story of Semmelweiss. From the "delivery position" theory, the logical prediction was: *If* you change to a lateral delivery *then* the fever deaths will decline. This observable consequence would have been true *if* the delivery position theory had been true. This prediction constituted an experiment which Semmelweiss could perform. From the "priest" theory it followed that: *If* you re-route the priest *then* the fever deaths will subside. Again there is a particular experiment which can be performed involving a result that would be expected *if* the tentative "priest" theory were true. Finally, from the "cadaveric matter" theory it followed by prediction that: *If* you wash hands thoroughly before labor room examinations, *then* the deaths will subside. Similar examples can be found in the story of Newton.

Let me illustrate the nature of prediction. It happened one day that my wife brought home a tempting package of beautiful plums. Because I consider it my moral obligation to consume all such delicacies as soon as humanly possible, it wasn't long before I had devoured the entire batch. In the process, however, I wanted to be sure I ate the sweetest plums first. (I've never had much sympathy with my wife's noble self-denying principle of saving to last the juiciest piece of pie or the tastiest piece of steak.) My problem then was to determine which plums were the sweet ones. My "keen powers of observation" and "active imagination" had led me to conclude by the end of the last plum that it was probably the dark plums which were sweetest. (You can see now that the prodigious gastronomical effort had really been all in the name of Science, Wisdom, and Truth.) My tentative theory was that "All Dark Plums are Sweet" (notice the general character of the theory). I believed it was the dark ones, not the soft ones or the oval ones which were sweet. Perhaps the analogy with dark sweet raspberries led me to this theory; I do not know. In any event, I felt it was my responsibility to fully test this theory. To this end, and at my bidding, my wife brought home a second batch of plums the next day. With the calm deliberation of a dedicated scientist at work, I immediately unwrapped the package, seized the first dark plum I saw, and made the following prediction: "This plum will be sweet."

Two important points emerge. First, predictions always apply to *new* particulars, never to the same particulars from which the tentative theory originated. If predictions are intended as part of the testing process for a theory, it makes no sense to predict something already known. In the example, it would have been ridiculous for me to devise my "dark sweet plum" theory on the basis of a batch of plums and then test my theory by predicting that the *same* batch will prove that "dark plums are sweet." That is like "pulling oneself into the air by one's bootstraps." It is circular reasoning and a useless procedure in science. My plum prediction, to be useful in testing, had to apply to a different batch of plums. The second point to emerge is that predictions follow from a theory by the logic of *deduction*.

b)   *Logic: Deductive* — Like abduction, deduction is a kind of reasoning or thought process. Unlike abduction, however, it *does* proceed according to a precise and rigid set of rules. Most of us are quite familiar with this kind of logic, without ever having taken a logic course, because we use it everyday in our own thinking. If dinner is served only between four and six p.m., and if I am hungry but have a two hour class beginning at five p.m., then I must eat between four and five.[22] Elementary, my dear Watson! It would almost seem that we are born with the rules of deduction stamped on our brains.

The force of deductive reasoning is in its *necessity*. From the premises "All men are mortal" and "Socrates is a man," we are forced to conclude that "Socrates is mortal." In any valid deduction, if the premises are true, the conclusion *must* be true.[23] This element of necessity applies then to predictions drawn from theories. These predictions are the "observable *consequences*" mentioned above. If one thing is a consequence of another, it follows from the other necessarily.[24] Experiments then can be seen to follow necessarily (by deduction) from theories. If the theory is true, then the prediction must be true. If "All Dark Plums are Sweet" is a true theory, then "This dark plum in my hand will be sweet" must also be true.

Deduction frequently is reasoning from general to particular, and in science, this is especially true. Theories like "All men are mortal" or "All dark plums are sweet" are clearly *general* while the prediction

"This dark plum in my hand will be sweet" is quite *particular*. That deduction and hence prediction go from general to particular is evident from Figure 3 above.

c) *Characteristic: Mechanical* — The rigid nature of the rules of deduction makes the process of prediction characteristically mechanical. At times it is so mechanical as to be almost automatic. One "automatically" sees what experiments must be done; checking the room next door, changing the priest's entrance, tasting the next dark plum. There are, of course, many situations where the necessary experiments are complex, requiring considerable ingenuity to devise and perform. There are also, of course, predictions which are so indirect that they must come through dozens of linked deductions from the theory. Often there are many additional (and even hidden) assumptions required, making the prediction a test of these premises as well as the theory. But the logical connection remains deductive and, in that sense, the predictive process is mechanical. It is as bound to the laws of deduction as a machine to the laws of physics. To participate in this stage of the scientific process requires precise, careful, analytic thought.

But no prediction can ever test a theory unless an experiment is performed. It may only be performed in the mind with full confidence in how nature would behave, as in Einstein's famous *gedanken* (thought) experiments, but some comparison of the prediction to the empirical world is essential. Prediction is incomplete—and with it the whole process of science—without the third and final stage of *confirmation*.

3. Confirmation

a) *Nature* — When a prediction has been made (an experiment suggested), it remains to be performed. It is here that science turns to the domain of sense experience as its ultimate criterion of

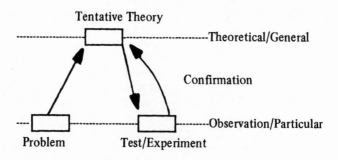

Figure 4: **Confirmation**

truth. The process of confirmation is that stage of the scientific method where the results "come in" on the predictions made. These results come from *particular* experiments, yet they have bearing on the truth of theories which are quite *general.* This impact of the particular on the general is illustrated in Figure 4.

b) *Logic and Character* — Experimental results are sometimes ambiguous and often open to various interpretations because of their complex derivation from various theories. Nevertheless, in principle, those results are of two types.

i.   If the experimental results do not agree with the predicted results, the theory is disconfirmed or falsified. Given a valid prediction and a carefully performed experiment it only takes one failure to justify rejecting a theory.[25] When Semmelweiss tried changing the delivery position and priest's entrance only to find no change in death rates, he immediately dismissed the respective theories from which these experiments were deduced. When upon entering and searching the room next door you do not see, hear, smell, touch, or taste your roommate, you naturally reject the theory that he or she is there. It only takes one sour dark plum to falsify the theory that "All dark plums are sweet." When the truth of a theory necessitates the truth of a prediction, and sense experience denies that prediction, we are forced to conclude that the theory is false.[26] The scientific enterprise must at this point return to its problem and to the process of discovery for consideration of an alternative working hypothesis.

ii.   If on the other hand, the experimental results agree with what was predicted, we say that the theory has been confirmed. When the expected result is obtained, this lends credibility to the theory. Seeing your roommate in the room next door certainly confirms your theory that he or she is there. Observing fewer and fewer deaths after instigating the hand washing procedure certainly confirmed Semmelweiss' theory of "cadaveric matter." But this is most definitely *not* to say that such a theory has been *proven.* If, for example, the first dark plum of the second batch turned out to be sweet, I'd be on precarious grounds proclaiming that my theory had been *proven* and that indeed "All dark plums are sweet." More of the reasons for this are discussed in Chapter Five below, but it is apparent here that the second dark plum may yet be sour, or the third, or the fourth, and so on. There may have been a sour dark plum in China in 1613, or one there now, for that matter. There may yet be a sour dark plum in 1990. The point is that while every successful prediction makes its theory progressively more certain—the process is never complete.[27]

The name given to this process of accumulating support for a general statement (here a theory) from the evidence of many particulars is *induction*. Like both abduction and deduction, induction is a kind of reasoning or thought process. Like abduction, it moves from particular to general. But induction differs from abduction because of induction's inherently *cyclic* character.[28] This is the character that makes it forever incomplete. Induction has been likened to a counting process whereby one adds confirmation on top of confirmation; always improving the probability of truth, but never finishing the sum. When another predicted experiment turns out as expected, the theory is further confirmed but the testing cycle must still be repeated with new predictions. This cyclic character of confirmation is illustrated along with the result of disconfirmation on the *now completed* diagram in Figure 5.

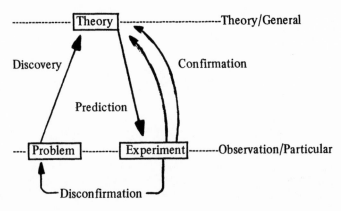

Figure 5: Scientific Method

In summary it must be observed that while it is fairly easy to disconfirm or falsify a working hypothesis, it is difficult if not impossible to fully confirm one. Lest any of you be concerned that this fact has doomed me to a life of perpetual plum popping, I can assure you that I have been spared that fate. The first dark plum of the second batch puckered me up like a persimmon. Scratch one theory and back to the drawing board!

SUMMARY

The features of the three process analysis of science as an activity can be summarized in the following chart.

Process	Process Logic	Process Nature	Process Character
1. Discovery	Abduction	Particular (Problem) to General (Theory)	Creative
2. Prediction	Deduction	General (Theory) to Particular (Experiment)	Mechanical
3. Confirmation	Induction	Particular (Experiment) to General (Theory)	Cyclic

Figure 6: Scientific Method

One of the best examples of the necessary characteristics of a good scientist is that king of detectives, Sherlock Holmes. Some like to remember Holmes with magnifying glass and tape measure busy in a darkened alley searching for clues that mere mortals might have overlooked. Others will remember him for those long periods during which Holmes, with pipe in hand, pondered over solutions. But of course Holmes did neither unless there was a problem at hand. No scientist gathers observations or searches for hypotheses unless there is a problem at hand. You may recall that, between cases, Holmes was disconsolate, finding solace only in the use of morphine and cocaine. The problems came sometimes in the form of obvious appeals from the bunglers at Scotland Yard. When a case escaped the framework of normal crime, these anomalies were frequently passed on to Holmes. At other times Holmes found his own problems. The active mind sees problems where most of us see only familiar objects. On one occasion Dr. Watson found Holmes using a lens and forceps to examine what Watson called ". . . a very seedy and disreputable hard-felt hat, much the worse for wear, and cracked in several places."[29] Holmes objected and said of it to Watson, "I beg that you will look upon it not as a battered billycock, but as an intellectual problem."[30] The hat led them, because of some almost unnoticeable peculiarities, to one of their most interesting adventures; but only because Holmes had seen a problem in it from the start.

Now Holmes often *said* of his "inductive" methods that one ought never to form an hypothesis until *all* the facts were in. Indeed such *is* the approach of the ants. "No data yet," said Holmes, "It is a capital mistake to theorize before you have all the evidence."[31] But here Holmes was quite mistaken. Of course, first theories are only tentative and rarely complete, but without some theorizing it is strictly impossible to collect all the relevant data. Holmes' *actions* were more on target. He always knew where he wanted to look for clues and which observations *were* relevant. This suggests that he did have preliminary hypotheses which were continuously updated as he collected additional information and pondered anew during long bouts with his pipe or violin.

Holmes' behavior illustrates that no scientist proceeds without interruption through the three processes outlined in this overly simplified discussion. In actual practice, the path is far less direct. In discovery, for example, the scientist, like Holmes, does not usually jump in one creative step from his problem to a fully developed answer. The discovery step may ordinarily consist of a series of subcycles *each* involving a preliminary hypothesis, predictions and tests. Initially broad hypotheses led Holmes to predict that certain clues might be found, to look for those clues, and finally either to pursue and narrow his hypothesis or to alter and move to another line of thought.

Sometimes the predictions could be tested in his head using information already known. At other times Holmes was forced out to check for facts or forced home to his lab to test a prediction there. Each subcycle is guided by its predecessor. When an hypothesis was confirmed, the subsequent cycle moved along the same line and became either more general or more specific as was required to make the initial problem explicable. If a preliminary hypothesis was disconfirmed, the subsequent cycle moved off in a new direction. Naturally many of these subcycles occur almost unconsciously. The hypotheses often arise creatively out of an analogy and are tested quickly against the experience of the scientist.[32] With this "try and see" approach, the scientist comes closer and closer to the full answer to his problem.

Of discovery and the process of abduction—or "reasoning backwards" as he would call it—Holmes had this to say:

> Most people if you describe a train of events to them will tell you what the end result would be. They can put those events together in their minds, and argue [deduce] from them that something will come to pass. There are few people, however, who if you told them a result would be able to evolve from

their own inner consciousness [abduction] what the steps were which led up to that result.[33]

We see in Holmes the two essential characteristics of the creative scientist. On the one hand, his attention to detail and preciseness of thought were famous. He sometimes dropped to his hands and knees and moved about like an ant, utterly pre-occupied in his observations. When examining a room on one occasion, Watson reported that,

. . . he whipped a tape measure and large round magnifying glass from his pocket. With these two implements he trotted noiselessly about the room, sometimes stopping, occasionally kneeling, and once lying flat upon his face. So engrossed was he with his occupation that he appeared to have forgotten our presence, for he chattered away to himself under his breath the whole time, keeping up a running fire of exclamations, groans, whistles, and little cries suggestive of encouragement and hope. As I watched him, I was irresistibly reminded of a pure-blooded, well trained foxhound as it dashes backward and forward through the covert, whining its eagerness, until it comes across the lost scent. For 20 minutes or more he continued his researches, measuring with the most exact care the distance between marks which were entirely invisible to me, and occasionally applying his tape to the walls in an equally incomprehensible manner. In one place he gathered up very carefully a little pile of gray dust from the floor and packed it away in an envelope. Finally he examined with his glass the word upon the wall, going over every letter of it with minute exactness. This done, he appeared to be satisfied, for he replaced his tape and his glass in his pocket.

"They say that genius is an infinite capacity for taking pains," he remarked with a smile. "It's a very bad definition, but it does apply to detective work."[34]

You will have to read "A Study In Scarlet" to see the usefulness of this precise, careful side of Holmes' personality.

On the other hand, Holmes was also a dreamer. His powerful imagination sought the simplest theories to answer the problem, regardless of how fantastic that theory might be. This "dreaming" was actually Holmes' way of applying creative imagination to the invention of likely working hypotheses which not only explained the facts of the problem but permitted testable predictions of other new facts. On one occasion, according to Dr. Watson after returning from collecting clues,

He took off his coat and waistcoat, put on a large blue dressing-gown, and then wandered about the room collecting pillows from his bed and cushions from the sofa and armchairs. With these he constructed a sort of Eastern divan upon which he perched himself cross-legged, with an ounce of shag tobacco and a box of matches laid out before him. In the dim light of the lamp I saw him sitting there, an old briar pipe between his lips, his eyes fixed vacantly upon the corner of the ceiling, the blue smoke curling up from him, silent, motionless, with the light shining upon his strong-set aquiline features. So he sat as I dropped off to sleep and so he sat when a sudden ejaculation caused me to wake up, and I found the summer sun shining into the apartment. The pipe was still between his lips, the smoke still curled upward, and the room was full of a dense tobacco haze, but nothing remained of the heap of shag which I had seen upon the previous night.[35]

I do not mean to suggest that creativity requires all scientists to crawl upon their bellies in the dust or sit up all night speculating. But this balance of attention to observation and imaginative thinking is essential in the process of science.

In conclusion, the activity of science is not some secret set of rules which when properly applied automatically and mechanically produces theories. It does involve a certain mechanical character in its deductive prediction stage, and it does include methodical attention to observations in seeing the problems and doing the experiments. But this is balanced by the need for a kind of imagination or intuition that together with these others makes the process creative. Non-scientists need not be intimidated by such a process.

CHAPTER TWO

The Development of
Scientific Ideas:
History of Science

INTRODUCTION

Review

The purpose of Chapter One was to investigate the nature of science when considered as a dynamic human activity. By examining the methods used in science we sought to confront certain misconceptions about the way scientific ideas are *formed*. These misconceptions contribute to the widening of a gap between scientists and those persons involved in the humanities. This gap of "two cultures," as C. P. Snow describes it, consists largely of a barrier in communication leading to mutual ignorance and to negative attitudes. As the first chapter revealed, a closer study of the formation of scientific ideas shows that a) it is not very different from common sense and b) it is not a magical mechanical process known only to an elite scientific "priesthood." Overcoming such misconceptions helps to narrow the "Snow Gap."

A Distinction

In this chapter we are once again concerned with science as an activity. But here we must confront a different misconception; a misconception that requires us first to make a distinction between the *formation* of scientific ideas and the *development* of scientific ideas. While both pertain to science as an activity, they differ in focus. The formation activity, as evident from Chapter One, concerns the scientific enterprise at the individual level. It describes the methods of inquiry used by the practicing scientist searching for generalizations to explain experience. In short it is a *psychological* activity. On the other hand the development activity, for our purposes, concerns the scientific enterprise on the group level, over a period of generations. It describes the life-cycle of a given scientific idea, born of discovery, confirmed, altered or modified, and ultimately discarded through progress in the history of science. In short, the development of scientific ideas is concerned with science as a *social* activity. The distinction might also be

31

expressed as the difference between microscopic activity and macroscopic activity.

A Misconception

The misconception that frequently arises concerning the development of scientific ideas is that there is a cummulative, uninterrupted straight-line progress in science from generation to generation. Many believe that there are no dead-ends in science; that each new idea builds on the last in an uncanny thrust towards truth.

Such a misconception is often caused by distorted histories of science, especially those found in science textbooks. Such histories neglect to show the mistakes and wrong turns science has taken. The influence of personal and social factors is often overlooked or minimized.[1] They are included only as background to the current theories presented in the text and thus are quite naturally distorted by a process of selective editing. As Thomas Kuhn, a well-known historian of science, puts it,

> Inevitably, the aim of such books is a persuasive and peda-
> gogic; a concept of science drawn from them is no more likely
> to fit the enterprise that produced them than an image of na-
> tional culture drawn from a tourist brochure or a language
> text.[2]

A common result of this misconception of scientific development is the further widening of C. P. Snow's gap. As the layman is misled regarding the history of scientific progress, he is intimidated by the apparent infallibility of the enterprise.

Preview

The natural solution to the misconception described is to return to a clear examination of the history of science. It is our purpose here to look at two examples in the history of science as a macrosocial activity in order to observe what general features *do* accurately apply to that process of development. The *main point* is that science develops in history neither in what we will call a strictly 'evolutionary' way nor in a strictly 'revolutionary' way, but rather by a process which combines both.

This chapter is not a history of science. Space limitations and present purposes make that impossible. The two examples chosen for investigation are first, parts of the history of scientific ideas of motion and second, parts of the history of the scientific ideas in astronomy. The choice of these two was based in part on their centrality in the works of historians of science and in part on their familiarity.[3]

32

THE DEVELOPMENT OF SCIENTIFIC IDEAS OF MOTION

Aristotle

The work of Aristotle in the fourth century B. C. was one of the earliest attempts at what we would today call science. Until his time there had not been much interest in the physical world as something worth studying in and for itself. Of course there had been Egyptians and Babylonians who became quite adept at manipulating the physical world and even at observation, but Aristotle added the necessary drive for generalization and abstraction that did much to further the cause of what was then known as natural philosophy. While many of his conclusions were later found to be quite wrong, his importance remains.

An obvious feature of the world that for Aristotle required explanation was the fact of motion. The *key* to understanding Aristotle's conclusions is that for Aristotle *motion was an unnatural phenomenon.* By this he meant that things naturally prefer to be at rest. Motion then, wherever it occurred, always required an explanation. Something in motion can be understood only by finding its mover. Objects at rest need no such explanation. [4]

For example, Aristotle determined that objects on earth which move by falling when unsupported, must have a mover. This mover, he said, is their internal desire to be at the center of the earth. Heavenly bodies which move in cyclical patterns are moved, said Aristotle, by their desire to copy the perfect circular motion of the Unmoved Mover.

Figure 1: **Pushing an Object: Apparent Support for Aristotle**

While such a theory seems humorous today and hardly scientific, we must understand that it arose from simple observations in a way which most of us would find hard to fault.

It seemed obvious to Aristotle that an object in motion remained in motion only so long as something was moving it. When the mover is taken away, the motion ceases.

So long as we push on something the object moves! When we rest, the object stops. It also seemed obvious that so long as the mover acted in a constant way, the motion involved would be constant. So long as we push with a constant effort, the object will roll at a constant speed. Thus Aristotle's idea that motion was unnatural, mistaken as it was, nevertheless was based on observation and probably could not have been otherwise given the observations of that day.

The Aristotelian view of motion, along with most of Aristotle's other work, was lost to civilization for almost a thousand years. For some reason the documents were misplaced and overlooked. It was not until the Middle Ages (*circa* 1200 A. D.) that they were re-discovered and translated first by Arab commentators and later by Catholic church leaders. Especially in the work of St. Thomas Aquinas, Aristotle came to be the basis for acceptable science in that day. St. Thomas' famous proofs for the existence of God are based to a large extent on Aristotle's theories, especially on his theory of motion.

But by the 14th century, problems began to arise with Aristotle's theory. In the first place was the problem of projectile motion.

Figure 2: **Shooting an Arrow : A Problem for Aristotle**

If, for example, the forward motion of an arrow should cease as soon as it loses contact with its mover in that direction, the bowstring, why does it not then drop immediately to the ground, moved only by its desire for the earth? Its arcing trajectory could not be explained by Aristotle's theory of motion. In the second place, it was finally noticed that falling bodies speed up as they fall. The longer they fall, the faster they fall. But if they are being moved by a constant desire for the earth, Aristotle's theory seemed to predict that they should fall at a constant speed.

Neither of these observations fit Aristotle's theory; they were anomalous observations. From Chapter One we recall that anomalies set the stage for discovery of new theories. They are the falsifying evidences which prompt the rejection of a theory.

But before any theory is rejected—and especially one as old and respected as was Aristotle's in the 14th century—there are always attempts to show how the anomalies can be fitted, after all, into the theory. Sometimes small modifications of the theory are necessary, but scientists wisely and yet dogmatically refrain from throwing out too much too soon. The 14th century Aristotelians explained the arrow's continued motion by pointing out that the air near the head of the arrow became compressed by the motion and subsequently rushes around to the tail to prevent a vacuum from forming and there serves to push the arrow forward.[5] The increasing speed was explained by

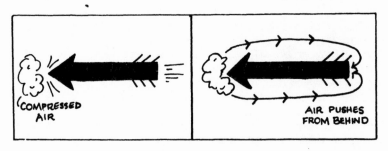

COMPRESSED
AIR

AIR PUSHES
FROM BEHIND

Figure 3: **An Attempted Answer to Aristotle's Problem**

pointing out that as an object falls, the air column above it increases and that below it decreases, producing an ever *increasing* push downward on the object causing its speed to increase.

But this patchwork rescue maneuver was ultimately unsuccessful. It stumbled on its own internal contradiction. How could the air through which the arrow flies be both its resistance and its propellant? Further, should not a thread tied to an arrow's tail blow

35

forward and not backward as experience reveals? Clearly the time for change had come. A step in the *development* of the scientific ideas of motion was necessary.[6]

Impetus Theorists

At about the time of the Aristotelian difficulties there arose a fairly consistent body of teaching which offered an alternative theory of motion. Set forth to a large extent by a group from Merton College, Oxford, this view is important because it is difficult to classify it as basically Aristotelian or as basically modern.[7] Its basic tenet was that when a body is put in motion by a mover, it is given a certain amount of "impetus" as well. This impetus continued to serve as the mover of the object, especially of projectiles, even when the original mover was no longer present. It appears that this impetus was supposed to be a mysterious fluid or thing inside the object itself.[8]

Clearly this view could explain the situations Aristotle's could not. Projectiles flew on in their trajectory in proportion to the impetus acquired from the mover, gradually losing their "inertial fluid" until they fell. Falling bodies actually accumulated impetus or "accidental gravity" as they fell causing them to increase in speed. It could perhaps be argued, however, that this view did not yet disagree with Aristotle's basic view that *motion is unnatural* and requires a mover to be explained.

Galileo

It was not until the 17th century that the scientific ideas of motion clearly changed. The century was a century of remarkable genius. It was lighted by more human creativity perhaps than any other century before or since. The Renaissance was under way and progress was not limited to science. In art, philosophy, music, and many other fields, change was the watchword.

Galileo is perhaps best remembered for something he probably never did; dropping balls from a leaning tower in Pisa.[9] Nevertheless his contribution in the areas of motion and astronomy were monumental. The birth of modern science is sometimes dated from the work of this man.

Galileo did something quite different from Aristotle. The *key* to his view of motion was that *motion is as natural as rest.* This revolutionary idea suggested that objects in motion *naturally* tend to continue in motion.[10] This eliminated immediately the need for any explanatory mover once an object was in motion. The old problem with projectiles simply disappeared; it was *natural* for the arrow to carry on in its trajectory once set in motion by the bowstring.[11] The importance of this change in the fundamental assumption regarding motion included the elimination of any need for mysterious fluids, or *spirits* that had hitherto kept science and religion tied very closely.

The difference between Aristotle and Galileo on motion is sometimes illustrated by the story of a chandelier.[12] It is said that Galileo was daydreaming during mass in a large Renaissance cathedral one day. A huge overhead chandelier happened to be swung by a draft of air. As he gazed, Galileo was caught up with its regular pendulum-like motion. The Aristotelian view of motion suggested that the chandelier was struggling to come to rest in its natural motionless state. But suddenly, it is said, Galileo saw that the same chandelier could just as easily be seen as struggling on each swing to swing as high as it had in its previous swing. It was prevented from doing so only by air resistance and mechanical interference in the pivot point. By a stroke of imagina-

tion and a sudden shift of perspective, the modern theory that *motion is as natural as rest* was born. While oversimplifying somewhat, this illustration does capture the change and contrast which characterized this important step in the development of scientific ideas of motion.

Newton

The classical modern statement of the scientific idea of motion is found in Newton's First Law of Motion. Paraphrased, that law states that "Bodies at rest tend to remain at rest, and bodies in motion tend to remain in motion, in a straight line, at a constant speed, unless acted upon by a net force." One may think of the Law of Inertia as a law of "status quo." Unless interfered with, a body will continue in the state of status quo; i.e., in motion if already in motion, or at rest if already at rest. This principle remains uncontested today.

Conclusion

While it is fresh in our mind it is well to make some general comments about what this brief history of motion tells us about the development of scientific ideas.

1. In the first place, the transition from Aristotelian to Galilean theories suggests that often there is need for radical changes of direction in the development of scientific ideas. Whether by a flash of intuition in the mind of an individual scientist or by the search for new frameworks and perspectives by many, there will be times when a fresh start must be made. The old views must be rejected as wrong or at least as misleading. This switch of perspective is referred to in modern psychology as a *gestalt* phenomenon. Most of us can think of examples in our own experience where an old problem takes on a new dimension, or more graphically, where a box seen from below suddenly appears as a box seen from above. This feature in the development of science is pre-

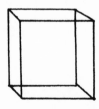 

Figure 4: Gestalt Pictures

cisely the same creativity seen to characterize the formation of scientific ideas during discovery. Often no new observations are necessary, as with Galileo's chandelier, only a fresh look at the old ones.

2. A second comment is that the development of problems or anomalies in a theory produces the kind of situation conducive to

this creative, radical thinking. The ever increasing cracks in the predominant Aristotelian views freed the minds of men like Galileo to think in fresh ways.

THE DEVELOPMENT OF SCIENTIFIC IDEAS IN ASTRONOMY

Aristotle

The second major historical example concerns the development of ideas about the structure of the universe. Here too we begin with Aristotle. Certainly there were many before him to set out theories of the universe. We begin with Aristotle only because of the profound influence he had on so many generations of subsequent thought in this area.

The *key* to Aristotle's theory of the universe was that every celestial body revolved around the earth in perfect circular motion, copying that of the Unmoved Mover and always conforming to Aristotle's theory of motion. Because the earth was at the center, Aristotle's theory is said to have been geo-centric. This geo-centricity strikes us today as primitive in its misconception. Once again, however, it is important to understand how this view arose quite naturally from observations which we today would still find hard to fault. It seemed quite obvious from the daily rising and setting of the sun and moon that they must rotate about the earth. The stars too because of their daily motions must move about the earth in a sphere of their own.

The idea that celestial bodies rotated about the earth attached to large transparent concentric spheres was not new to Aristotle. Both earlier and subsequent astronomers held this view. Earlier views limited the number of spheres necessary to account for celestial motions to two only. Dante's theory incorporated ten such distinct spheres. To provide a more accurate accounting of precisely observed celestial motions, Aristotle's system incorporated between 45 and 55 spheres, depending on how one interprets Aristotle.

There was a sphere on whose surface lay all the stars and which rotated every twenty-four hours on an axis connecting the earth's center and the North Star. The sun's sphere rotated daily on an axis called the 'ecliptic,' and gained a full revolution each year against the star sphere. The planets each had spheres with unique axes and periods varying slightly from twenty-four hours. It was Aristotle's view that each sphere was moved by the next larger sphere all the way out to the Unmoved Mover whose perfect circular motion was thus the ultimate explanation for all motion in the universe.

Aristotle's theory ran into difficulty only when improved observations revealed that simple circular motion could not account for the *retrogression* of certain celestial bodies. *Retrograde motion* is observed when sightings of a planet are made on consecutive days at identical times. At first the planet appears to move consistently in one direction for successive sightings. Its apparent position then remains nearly fixed for several days, and then it appears to shift back in the direction from which it came before stopping again and finally resuming its original direction of shift. If planets simply circled the earth on uni-

RETROGRESSION: (All sightings at midnight)

Figure 5: **Retrograde Motion**

formly moving spheres this erratic behavior could not be explained. The anomalous retrograde motion led to the disconfirmation of Aristotle and the birth of a new theory.

Ptolemy

Ptolemy lived about 500 years after Aristotle, during the Hellenistic period in the Egyptian city of Alexandria. Ptolemy retained both Aristotle's concept of geo-centricity and his assumption that celestial motion can only be in perfect circles. But Ptolemy was able to account for observed irregularities from such circularity, especially retrogression, by assuming that the motions of celestial bodies are compounded of many circular motions, one on top of the other. The resultant motion is not circular, but it is *reducible* to circular motions.

Ptolemy employed basically four theoretical devices to accomplish this reduction to circular motion. The first two, the *deferent* and the *epicycle,* go together and are the easiest to understand. The heavenly body moves in a circle (epicycle) about an imaginary point in space which itself is moving in a circle (deferent) about a second point. Like swinging a ball around your head (epicycle) while riding a fast merry-go-round (deferent), the compound motion of the ball is quite complex but clearly not a perfect circle. A third device used by Ptolemy is called the *eccentric.* As the name suggests, it means something off-center. The centers of the celestial planetary deferents were supposed to be the center of the earth in a strictly geo-centric theory. But to account for

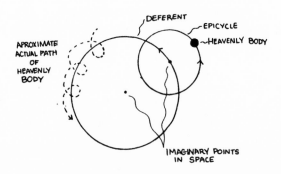

Figure 6: Deferent and Epicycle

observations Ptolemy was forced to make the centers of these deferent orbits slightly *off* earth's center, i.e., they were eccentric. The fourth

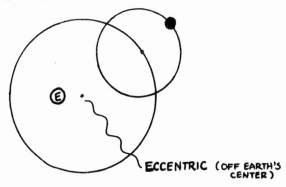

Figure 7: Eccentrics

device is perhaps most difficult to understand and historically the most controversial. The *equant* was a device which basically meant that heavenly bodies did not move in their epicycles at a uniform speed. At some times they would speed up and at others slow down in their perfectly circular epicycles.

Although there were only these four basic *types* of devices,

41

Ptolemy was forced to use one or another of them in about 80 different places in his system in order to account accurately for the astronomical observations of his day.

It was over 1400 years before the basic Ptolemaic system began to encounter serious problems. Naturally there were modifications during that time to accommodate improved observations, but no major change. The problem that did eventually surface did not lie in the explanatory power of the theory. Even after Copernicus presented his alternative, the Ptolemaic theory was more than adequate to account for all observations. The basic problem, to which Copernicus objected, was that Ptolemy's use of the equant violated Aristotle's old prejudice towards perfect circular motion. Perfect circular motion meant circular motion at a *constant speed*, but Ptolemy's equant denied this for celestial bodies. Ironically, it was his deep *personal* commitment to a quasi-religious notion of circular motion that prompted Copernicus to propose his radical contribution to the development of scientific ideas of the universe.[13]

Copernicus

Copernicus lived in the late fifteenth and early sixteenth centuries. The *key* to his theory was of course that the sun was at the center of the universe and that the earth spins on its own axis. Despite the apparently revolutionary character of his theory, Copernicus apparently never intended any break whatsoever with the basically Aristotelian assumptions of Ptolemy.[14] It is quite possible that Copernicus saw his helio-centricity not as a description of the way heavenly bodies are actually arranged, but simply as a way of imagining things in order to simplify calculations of planetary and stellar motions. This interpretation is all the more likely when seen in the context of Copernicus' conservative Aristotelian motive for rejecting Ptolemy's equants. In fact his helio-centric view did not create any serious controversy until Galileo came along because it was generally regarded as just a convenient calculating device and not as a statement about where the sun is *actually* located.

Copernicus regained the Aristotelian ideal of *uniform* circular motion without needing the equant. But he still needed the epicycles, deferents, and eccentrics of Ptolemy to account for observations, despite his radical rearrangement of the universe. Copernicus did manage to reduce the number of such devices necessary to between 34 and 48, compared to Ptolemy's 80. This leads us to make a more direct comparison of what appear, from our vantage point in history, to be two radically different theories.

There were at least two consideration in *support* of the Copernican model. First, the Copernican theory was clearly a more *elegant* system insofar as it reduced the number of devices necessary to explain the heavenly motions from 80 to 48. Putting the sun at the center made the previously troublesome retrograde motion a natural consequence of the different orbital speeds of the planets. Second, in addition to this elegant simplicity, the Copernican system obviously did *not* require, as did the Ptolemaic system, that the stars and sun all revolve so fast (i.e., every twenty-four hours) about the earth. A simple spin of the earth in the Copernican system accomplished the same purpose.[15]

On the other side, however, there seemed to be far more evidence *against* the Copernican helio-centric theory. In the first place, and perhaps most important, the Copernican theory offered no better predictive/explanatory power than the Ptolemaic geo-centric system. In other words, both theories did comparable jobs of accounting for astronomical observations of the day. Second, the Copernican view challenged a theory that had in its essential features, been accepted and established for 1400 years. To us this may seem a small objection for Copernicus to overcome, but taken together with the first item it is overwhelming. It would be as if persons today suggested that gravity was really a repulsive force only appearing to attract things, but they accounted for experience no better than our present view of gravity. We would naturally feel that, unless the new alternative did a better job of explaining and predicting, it was not worth giving up the common sense view we have held for so long.

To make matters even worse for Copernicus, his theory *seemed* to fly in the face of the Christian Scriptures. In the Old Testament book of Joshua, Chapter Ten, verses 12 and 13, the reference suggests that the "sun stood still." In the day of Copernicus, this could only mean that the sun ordinarily moved. Had Copernicus actually been right about the earth moving around the sun, the Scripture would have indicated that the *earth* stood still. Galileo, a Copernican supporter, was later to feel the full force of this objection in his confrontation with the Church.

In addition to these first three objections, there were at least three more technical reasons why the Copernican theory was unacceptable. First, it still depended on the use of epicycles. The elliptical orbits necessary today to eliminate these epicycles did not come until the time of Kepler, and besides, they would have violated Copernicus' preference for perfect circular motion. Second, Copernicus still used the eccentric. Most of the planets were required to circle a point that did not coincide with the center of the sun. So even Copernicus' helio-centricity was not exactly helio-centric. Finally, the helio-centric system of Copernicus predicted that at various points during its yearly orbit about the sun, the earth's perspective or angle on the stars would change because of earth's constantly changing position. This *parallax* effect seemed to be a necessary and testable consequence of the Copernican view. But when astronomers searched for Copernicus' pre-

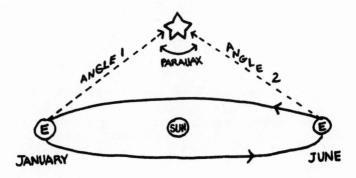

Figure 8: Parallax

dicted parallax, it simply was not there. The only reasonable conclusion to draw from this falsifying experiment and from the other overwhelmingly negative evidence, was that, despite its slightly greater simplicity, the Copernican theory was simply wrong in suggesting that the sun is at the center of things.

It is important to see that Copernicus did not convince the world. Even if he did mean that the sun was *really* at the center, his conservativism is revealed by his dogmatic allegiance to the Aristotelian theory of constant circular motion, by his retention of the Ptolemaic devices, and also perhaps most significantly, by the absence of controversy surrounding his work. The revolution in astronomy was by no means complete.

Galileo

The dispute in astronomy was brought to a focus mostly by Galileo. The question of helio-centric versus geo-centric theories became a real subject of controversy only with his prodding. There are at least three factors which brought this about. The first, and perhaps most important single factor, was Galileo's personality. Galileo was apparently an idealistic, argumentative, even perhaps stubborn and rash Italian professor. Without such a character, the advance in astronomy might have been postponed many, many years. It was this personality that perhaps prompted Galileo to risk confrontation with the Church to make an issue of something Copernicus had left unresolved.

A second, related factor is that Galileo was attacking not just the geo-centric Ptolemaic system but the entire framework of Aristotelian world-view pervading the civilized western world of the day. As we have seen, even Copernicus himself subscribed to the Aristotelian view of motion. For Galileo then, Aristotle was the real target. His theory of the universe was just as mistaken as his misguided theory of motion. It took the genius of Galileo to see the interconnection of problems in astronomy and motion and to trace them to their real source in Aristotle. The time had come for revolution.

The third factor precipitating the crisis of Galileo's day was the emergence of new astronomical data *apparently* supporting the Copernican theory.[16] The data was questionable at the time, because it had been obtained through use of a mysterious new device called the telescope. Most men of authority suspected its users of sorcery and witchcraft or just plain trickery. Nevertheless, Galileo accepted its use and the new observations it provided. (The attack on Galileo's support and use of this new device even came from the pulpit. A Dominican priest named Caccuni, is said to have preached against Galileo in Florence, using for his text, the Ascension passage "Ye men of Galilee, why stand ye gazing up into Heaven." The play on Galileo's name was hardly given in a humorous spirit.)[17] Among these new data were the first time observations of sunspots, the phases of the planet Venus, and the moons of Jupiter. For various reasons each of these provided good confirming evidence for the Copernican theory. Taken together, these three factors set the stage for Galileo to take on the defense of the Copernican theory before the established Church.

The battle with the Church, which was the political, academic, and spiritual authority in that day, is fairly well known. The Church was willing to accept the Copernican theory as a *calculating* theory, but *not* as a statement of the way the sun and earth really move.[18] Galileo on the other hand insisted that the sun was *really* the center and that the earth *really* moved.

In his defense Galileo tried virtually every trick in the book. With regard to the problem of the sun standing still for Joshua, Galileo showed that a literal interpretation of the story on the Ptolemaic system would have *shortened* the day instead of lengthening it and would have wreaked havoc with all the planetary relationships had the sun alone stopped moving, Instead, Galileo reasons, *all* heavenly motions must have stopped to give Joshua time to win his battle; and in particular the passage states that the sun stood still "in the midst of the heavens." This could only mean in the center of the system, argued Galileo, since if it meant only "in the middle of the sky," i.e., at noon, Joshua would have had no need for stopping it at all since there was plenty of daylight remaining in which to fight. Unfortunately, as miracles go, the Ptolemaic one would have been far easier to accomplish than the Galilean-Copernican one. As Galileo of all people should have known, to stop the earth's spin and motion would have flattened mountains and destroyed the earth. The effect of inertia is bad enough when horse-coaches strike a wall or ships run aground.[19]

Another argument Galileo tried to use was that the earth's spin and revolution about the sun could account for our tides. Unfortunately as we see from our vantage point today, it would have failed miserably to square with the facts and violated Galileo's own views of motion.[20] In a scramble to defend his view, Galileo seemed to be grabbing at straws.

In perhaps his most slippery move, Galileo tried to shift the burden of proof to the Church. He argued that before a theory could be condemned, it should be *proven* false. Naturally this must be undertaken by those who deny it, (in this case, the Church) because clearly they are more likely to find its faults than those who support it (in this case, Galileo). Galileo sought to shift the burden of proof basically because he had no proof for the Copernican theory which was acceptable to Church authorities.[21]

Despite its dismal state, Galileo carried on the battle. It became a matter of prestige. He had committed himself to the Copernican view and anyone who disagreed was belittling his authority as the foremost scholar of his day.[22] The personal stakes were high. In 1616, he was warned to stop defending Copernicus. In 1633 at the age of 69, Galileo was tried by the Church, threatened with excommunication, and forced to recant. Contrary to legend, he did not retort under his breath 'eppur si muove' (and yet it does move).[23] He was comfortably imprisoned for nine years during which time he wrote his most famous works and died in 1642, the year in which Newton was born.

Although Galileo had been defeated, the helio-centric theory had caught on, along with the modern view of motion. It was not long before the work of Kepler in astronomy and Newton on motion vindicated the struggle that Galileo had fought.

Conclusion

As in the case of motion, it is helpful to ask what this brief history suggests about the development of scientific ideas. It is evident that progress in science means choosing on occasion between competing, alternative theories. From what we have seen of the Ptolemaic/Copernican controversy, this choice involves several criteria.

In the first place, a theory is preferred if it shows better explanatory/predictive power. In other words, the theory which best fits the facts is the better theory. The Ptolemaic system with its four types of devices far surpassed the Aristotelian system in accounting for retrogression, hence the Ptolemaic Theory prevailed. But sometimes two alternative theories are equally good at explaining and predicting. Such

was the case with the Ptolemaic and Copernican systems. In this situation additional criteria are necessary for a choice to be made and thus for science to develop. What is, perhaps, surprising is that more often than not these situations are the rule and not the exception in the history of science.

A second criterion then is to choose the theory which is elegant (*e.g.*, shows greater simplicity). This quality is clearly open to interpretation depending on widely varying personal and social factors influencing the individual scientists involved in the issue. Elegance may for example only be what is familiar. It generally refers to theories with fewer initial assumptions and thus to theories which require the least amount of ad hoc bending, stretching, and patching. From our brief history we can see that this superior elegance was one of the few things contributing support to the Copernican theory. It required fewer devices than the Ptolemaic system to give a comparable accounting for the facts observed.

A third criterion is based on which theory shows a better "consensus" of support. In other words the theory which most authorities accept is the better theory. This kind of thinking is the one most difficult for the non-scientist to accept about science. It seems unthinkable that scientists, of all people, should in a sense, vote on what is true about the world. Yet this criterion perhaps more than any other single factor shapes the development of scientific ideas. In the confrontation between the Church and Galileo, the consensus was still not with Galileo. But before long, the tide had turned and helio-centricity prevailed. The same kind of process occurs over and over again in science even today as alternative theories of the universe's origin and subatomic structures are bandied about.

Naturally the consensus depends to some extent on other criteria, such as the first two above. But in an important sense the controversy which lies at the heart of scientific development is not a conflict of objective reason versus dogmatic faith, but rather a conflict of one set of assumptions against another set of assumptions in an arena of powerful personal and social factors. Galileo's dogmatic personal struggle to defend his prestige and the Church's equally dogmatic fight to defend its established position is just a case in point. In short, the choice of alternative theories as scientific ideas develop is rarely a clear-cut objective choice but involves significant personal and social factors.

THE DEVELOPMENT OF SCIENTIFIC IDEAS: EVOLUTIONARY OR REVOLUTIONARY?

Question

The purpose of this chapter has been to confront a misconception concerning the development of scientific ideas as a macro-social activity. That misconception is that science is a *strictly* evolutionary process. It is a serious misconception because it contributes to the widening of C. P. Snow's cultural gap, and to mistaken self-images by scientists. We have examined here two illustrations from the history of science in order to determine what general features characterize scientific development.

On the one hand, the *evolutionary* interpretation of science emphasizes the *continuity* of successive theories in history. The development of science is thought to be a cummulative, uninterrupted, straight-line process. Successive generations build upon predecessors by adding to the treasury of facts. It is almost unthinkable that science must ever backtrack. There are no blind alleys for science. This kind of thinking is sometimes illustrated by Newton's purported statement that he reached as high as he did only by standing on the shoulders of giants. Thomas Kuhn describes this view of science saying that,

Scientific development becomes the piecemeal process by which items have been added singly and in combination, to the ever growing stockpile that constitutes scientific technique and knowledge.[24]

On the other hand, the *revolutionary* view of scientific development emphasizes *discontinuity* in the progress of science. It suggests that science proceeds along a zig-zag course, with specific periods of crisis during which the course of thought is altered radically. At such revolutionary points, previous scientific theories are seen to have been false, and a new direction or approach is taken.

The contrast between these two views is heightened by the implications that each holds for our understanding of what we might call 'outdated science.' If, for example, we adopt the revolutionary perspective on scientific development, we must conclude that Aristotle's theory of motion is not science but merely a once popular myth. But then we are faced with the dilemma that the same observational methods used by modern science have been used to produce mere myths. What guarantee do we have that our present theories are any less mythical? If, on the other hand, we adopt the evolutionary approach, we must conclude that Aristotle's theory of motion was and is scientific. Yet this too seems paradoxical for it suggests that science may contain contradictory beliefs, i.e., Aristotle's and Newton's theories of motion. What evidence do we see from the examples discussed for holding either the evolutionary or revolutionary approach?

Evidence
1. Evolutionary

General — One general reason often cited in support of the evolutionary view of scientific development is the impression given by many textbook histories of science. It appears in such histories that every scientist knows just where he is headed, which facts are crucial, and what the result would be. But we have already seen that such histories may distort, perhaps not intentionally, but for "persuasive and pedagogic purposes."

Another general cause for holding science to be evolutionary is based on the mistaken view that in the formation of scientific ideas, discovery is a mechanical, automatic process. When individual theories are thought to be discovered by a process of fact gathering, it is natural to suppose that scientific development from theory to theory occurs in the same way. But we have already seen such a view of discovery to be misguided. We have referred to such a misconception of discovery as the Christopher Columbus Fallacy.

Anticipations — Perhaps the most powerful evidence for an

evolutionary interpretation of scientific development comes from historical research which reveals that most important scientific theories have been anticipated by earlier generations. This stresses the continuity and evolution of such theories. In the case of motion we have seen how the Theory of Inertia fully expressed by Newton was introduced by Galileo and anticipated before Galileo by the Impetus Theorists in Paris in the 1400's. In astronomy we could point to full-blown helio-centric theories, anticipating Copernicus, by Nicolas of Cusa in the Middle Ages, and as far back as Aristarchus, a Greek in 300 B. C. Some object that what may appear in retrospect as an anticipation of later ideas, is only evidence of an overly active historical imagination. Nevertheless the evidencè remains quite powerful, especially when obtained by careful historical methods.

Modifications — Another reason offered for the evolutionary view is that scientific ideas are frequently altered and modified as they come down from generation to generation. This suggests that later ideas evolve from earlier ones in a process very much like that of natural selection. The "environmental" circumstances of new observations push and prod the theories through minor alterations that accumulate. In the case of motion we saw the modification of Aristotle by the Impetus Theorists, and on some interpretations even the modificaion of the Impetus views by Galileo. Modification is far more clearly evident in the developing views of astronomy. Ptolemy modified Aristotle by adding the four devices discussed. Ptolemy was repeatedly modified by fourteen generations to accommodate improved observations. Among those modifications were those of Herakleides and Tycho. Some of these modifications even came close to the original Copernican system in terms of elegance by reducing Ptolemy's eighty devices to forty. Copernicus' theory was modified by Kepler with the addition of elliptical orbits and elimination of all remaining Ptolemaic devices. Such evidence certainly suggests an evolutionary march of gradual progress and development in scientific ideas.

2. Revolutionary

But there seem to be equally convincing arguments for a revolutionary view of scientific development. In the area of motion, the transition from Aristotelian to Galilean physics seems to have been sufficiently radical to call it revolutionary. This is so, it is argued, not only because the change was large, but because it seemed to involve the *rejection* of what had come before as *incompatible* with what was now held. That the Aristotelian and Galilean views of motion can be seen as incompatible could be illustrated by the chandelier story. The shift of perspective (gestalt) makes it revolutionary.

In the area of astronomy the contrast of Ptolemy and Copernicus is also cited as evidence of a scientific revolution. A geo-centric and a helio-centric system are obviously *incompatible,* hence the development from one to the other is revolutionary. What occurred by the end of the controversy was not the victory of a superior explanatory system over another but a shift of perspective. If the incompatibility of new and old is at the heart of the meaning of revolution, then Copernicus does seem to have wrought a revolution. Yet one must wonder even about that when one recalls his conservative motives.

Sometimes it is said that the transition from Newtonian to Relativistic physics is yet another example of scientific revolution. The "evolutionists" reply that Newton is just the low speed application of Relativity. The "revolutionists" reply that what Newton and Einstein meant by mass, force, length, and time were utterly incompatible. And so the dispute goes on.[25]

Conclusion: Two Versions of Balance

1. With apparently conflicting evidence on both sides of the question one wonders if any general conclusions can be drawn after all about the development of scientific ideas. Thomas Kuhn says that they can, and that scientific development is clearly *both* evolutionary and revolutionary in character.[26] He explains this by arguing that when science is viewed as a macro-social activity, it consists of two distinct parts.

On the one hand, says Kuhn, most science is what we may call *Normal Science.* Normal Science consists of the working out and testing of the currently accepted theory or framework. In Kuhn's words, Normal Science is "puzzle solving" in the context of the currently accepted scientific "paradigm." Just as Ptolemaic astronomers worked out, modified, and applied the basic Ptolemaic system for fourteen centuries, so today physicists are extending and applying the basic relativistic paradigm. This part of scientific development is clearly evolutionary in nature.

On the other hand, says Kuhn, there are times when problems with a theory persist. These anomalies force scientists to continuously patch and re-patch the theory to make it fit the facts. Adherents to the theory become defensive and usually dogmatic, struggling to preserve the theory's credibility. This dogma among scientists has the important positive function of preventing premature overthrow of existing theories.[27] Other scientists begin to seek and propose radical alternatives. This latter group, says Kuhn, is doing what we may call *Extraordinary Science.* Among the alternatives proposed, one eventually gains a consensus, and prevails. The discipline, be it physics, biology, or any other, changes directions, the new theory becomes the new paradigm, and a revolution has been effected. Normal Science is resumed. Examples of these revolutions, which occur infrequently in science might include the Copernican, the Galilean, the Relativistic, the Quantum Mechanical, and in biology the Darwinian revolutions. Naturally in other branches of science there would be others.

The point, says Kuhn, is that the natural development of scientific ideas involves both Normal and Extraordinary science at *different* times. It is, at different times, both evolutionary and revolutionary in character. This might be illustrated graphically as below.

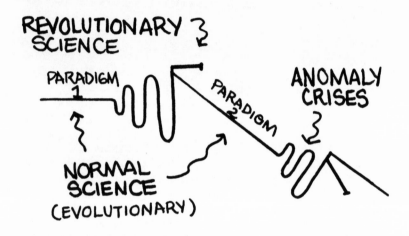

Figure 9: Scientific Development

2. It would be wrong, however, to assume that all historians of science agree completely on this view of scientific development. While Kuhn's view shows a balance of evolutionary and revolutionary activity, it focuses attention on the revolutionary periods of crisis and radical change. Stephen Toulmin presents an alternative view which retains something of the same balance but emphasizes instead the overall evolutionary nature of scientific development. The view is based on the model of organic evolution in biology.

Darwin's theory of evolution rests on several assumptions including 1) a natural variation of characteristics within any species and 2) the operation of a natural selection process whereby some variations are preserved and others eliminated.[28] Without these assumptions the evolutionary theory is no good.

As an example of biological evolution, let us suppose that giraffes once had shorter necks. Given an environment with only limited food supplies, the giraffes would compete with one another to exist. The first evolutionary assumption is that among the giraffes some would—by chance—happen to be a little taller than others. The second assumption then explains that these taller giraffes are more "fit to survive" in an environment where food on low branches becomes scarce. *Many* giraffes can reach it there while only tall ones have access to higher branches. Eventually this environment will "naturally select" the taller giraffes whose offspring will then be taller and so on.

Likewise, says Toulmin, in scientific development there are always 1) various competing theories to explain a phenomenon and 2) social and historical (i.e., environmental) factors which "select" the best variation as the preferred theory for that phenomenon and for that time.[29] In Toulmin's words,

> Science develops . . . as the outcome of a double process: at each stage a pool of competing intellectual variants is in circulation, and in each generation a selection process is going on, by which certain of these variants are accepted and incorporated into the science concerned, to be passed on to the next generation of workers as integral elements of the tradition."[30]

An important difference between Toulmin's views and that of Kuhn above lies in Toulmin's insistence on the need for competing intellectual variants "at *each stage*" of scientific development.[31] Kuhn argues that during normal science only one theory or paradigm is present. Toulmin insists that there must always be alternative theories under consideration. Toulmin's biological model and rejection of absolute scientific revolutions is the basis for his evolutionary emphasis.

But Toulmin's view is not a strictly evolutionary view either. It shows something of the same balance found in Kuhn's view. In the first place Toulmin, like Kuhn, rejects the idea that science develops in a "straight line" without false starts and blind alleys. In the second place Toulmin argues for the presence of both continuous and discontinuous factors in development. He says that scientific development "will then display *both* elements of continuity and elements of variability."[32] Toulmin's balance of evolutionary and revolutionary factors can be thought of as a constant or static balance while Kuhn's balance is more like the dynamic balance of a pendulum as it swings back and forth from one extreme to the other. The difference is significant but the similarity must not be overlooked.[33]

From our examination of Kuhn and Toulmin, as well as from the historical examples we have discussed, we can see that the development of scientific ideas is far from a cut and dried automatic process. It is a macro-social activity where personal and social factors figure largely. Science becomes a very human enterprise, not so different from the humanities. It is certainly not a magical process on this historical scale any more than it was not a magical process on the individual psychological level where ideas are formed.

CHAPTER THREE
The Nature of
Scientific Ideas:
Theory Structure

INTRODUCTION

In Chapter One we saw that science can be viewed from either a dynamic or a static perspective. The dynamic perspective treated in that chapter and in Chapter Two reveals science as both a microscopic and a macroscopic or social-historical activity. To study that activity is to study how scientific ideas are *formed* and *developed*. But what is revealed when we look at science from a static perspective? Having toured a factory and examined its operation, it is natural to ask about the nature of its products and about how these products are used.

Describing *how* scientific ideas are used seems to presuppose an understanding of *what* those ideas are. So before probing the use of science's products, we must be sure we know something of what the products really are. Hence these two questions focus the discussion of this and the next chapter respectively: "What are science's products?" and "How are they used?"

The *main point* of this chapter is that the products of science are several types of conceptual generalizations about experience, the meaning of which can be interpreted realistically or instrumentally. In the first two sections we investigate the various types of conceptual generalizations mentioned and the ways by which they are distinguished. In the last section we must turn to the question of how these generalizations—and especially scientific theories—are to be interpreted. The three views examined there are not exhaustive but are at least representative of the major alternatives.

GENERAL CHARACTERISTICS OF THEORIES

In a sense one could say that the products of science are found in the material fruit of technology. While this is certainly true, it is not our concern here. Technology is better understood as the application of science and so its fruit is only indirectly the product of science itself.

The purpose of science, as described in Chapter One, is to produce "conceptual generalizations about empirical experience." Basically, the products of science are theories. But what is the nature of these theories?

The phrase "conceptual generalizations" goes a long way towards clarifying the nature of theories. From Chapter One we recall that theories are found on that level of concern which is both *abstract* and *general.* [1] Abstractions always eliminate detail, and attempt to draw out selected features of that from which they are abstracted. Scientific theories then are abstract in the sense that they draw out intelligible features of experiences and thereby eliminate details. Corporate abstracts or legal "abstracts" do this well, and perhaps abstract paintings do, too.

The detail eliminated in the case of scientific theories is that detail associated with the *particularity* of the experiences in question. Theories are general, and so depending on whether the theory is of physics, or biology, or some other concern, the particularity eliminated is that detail associated with what makes this *specific* ball behave as it does, or this *specific* cell reproduce as it does, and so on. Generalizations ignore this particularity and focus instead on the common (or general) features of motion or reproduction.

In addition to being abstract and general, scientific theories are chiefly descriptive in nature. The elimination of detail and focus on common features is always to the end of *describing* the experiences in question.

Newton's theory of gravity is abstract in the sense that it ignores, for example, the shape or color of falling bodies. It is general in the sense that it is only indirectly concerned with why Tommy's ball fell as fast as it did or with why you weigh as much as you do. Finally, the theory of gravity is descriptive in the sense that it tells us only *how* falling objects or heavenly bodies behave. It does not tell us *why* they behave as they do. (More of this in Chapter Four.) Nor does it tell us that such objects *must* behave as they do. It is a serious mistake to suppose that scientific descriptions are really prescriptions for nature. Traffic laws tell us what we *must* do and yet quite frequently do not describe what people *actually* do. Quite the opposite, scientific laws only tell us what nature *actually* does and never what nature *must* do. We must not be misled into thinking that nature is "bound" in some way to the scientific "laws" which man has abstracted from her. The point is *not* that nature is likely suddenly to begin behaving contrary to man's descriptions, but only that "laws" of nature have a character very different from laws of society.

DISTINCTIONS: WHAT ARE SCIENTIFIC IDEAS?

Formulae, Laws and Theories

So far we have determined that the products of science are *abstract, general, descriptions* of experience. We have referred to these as theories. But are there not different types of scientific ideas? How does one sort out the meaning of such terms as theory, hypothesis, law, and formula? They often seem to be used interchangeably. While this interchangeability is quite acceptable for most elementary purposes—and thus for most of our purposes here—they can be distinguished. Unfortunately there is no single acceptable way to do this, but let me suggest two general approaches.

One approach is to distinguish some of these terms by their degree of confirmation. When, in doing science, a scientist discovers what may be an answer to his problem, the tentative answer may be referred to as an hypothesis. After a reasonable amount of testing and successful confirmation, the hypothesis may "graduate" to the status of theory. Finally, after years of consistent confirmation, it may be called a law. This system seems to fit such examples as the extremely well confirmed Law of Gravity or Snell's Law of Refraction and the less confirmed Theory of Relativity. Unfortunately, there are counterexamples such as the very well confirmed Kinetic Theory of Heat, which cause us to doubt the usefulness of this approach.

A second, and more widely accepted approach, is to distinguish among such terms by their degree of generality. Distinguishing the various types of scientific ideas depends on how far removed they are from the level of particular observations. At the lowest level of generality are the simple equations or formulae derived from tinkering with repeated observations. Examples would include the Balmer, Lyman, Brackett, Paschen and Pfund formulae for calculating atomic spectra wavelengths. These men "played" with observations until they found equations from which the observations could be derived systematically. This is not unlike the devices many people invent to help themselves remember a telephone number or the birthdays of one's family members.

The next highest level might contain laws. The distinction of laws from formulae is vague and this is illustrated by the Kepler's "Laws" which are little different from the formulae above, having been abstracted from the massive astronomical data accumulated by Kepler's mentor, Tycho Brahe. Certainly Newton's famous Law, $F = ma$, is also at least a very general formula for calculating forces, masses, or accelerations. But there is another important difference apparent between formulae and each of Newton's laws. These laws are not only more general but they include and use concepts which are not directly observable. These *theoretical terms* or *theoretical concepts* further distinguish laws from mere formulae. In the second law, $F = ma$, the concept of force is actually redefined in a new and technical way. In the first and third laws of motion the concepts of inertia and action likewise become theoretical terms. Other examples are many of the concepts from atomic physics such as nucleus, orbital electron, energy level, and neutrino.

An important question regarding such theoretical concepts is whether they refer to entities which actually exist, or whether they are merely useful concepts with which to understand a set of experiences. This is an issue which scientists themselves are often not too concerned to resolve. To them it would make little difference to the *usefulness* of atomic theory if electrons "really exist" or are merely convenient mental constructs. We will examine this issue more closely in the third section below on interpretation.

Finally, there are the most general ideas called theories. Because of their generality they "explain" (in a sense to be discussed in Chapter Four) a wide variety of apparently different phenomena. Newton's theory of gravity explained not only the effects of the earth on objects near it (as specified in part by the laws of motion) but also the effect of all heavenly bodies on one another and on the earth. Theories may then be understood to contain a group of laws connected to one another in a framework. This suggests that theories not only involve theoretical terms but also certain structuring principles as well. These principles are of two types: internal and bridge principles.

The internal principles include the bare, uninterpreted logical forms of the theory's laws, which relate theoretical terms.[2] But an internal principle may also be a statement relating the theoretical terms found in different laws. One law may be useful in calculating the measurements of one theoretical term, say, for example, the resistance of an electric circuit. A second law may describe how a second theoretical term, say power, can be measured in terms of voltage. An internal principle might later emerge which connects voltage and resistance and hence connects the two laws. The result is a single system (circuit theory) which has more applications than did either law alone. Clearly, it is the presence of these internal principles which gives a theory its explanatory power and its breadth of application. Since it is a central purpose of science to provide ever more inclusive accounts of experience, it is plain that the discovery of internal principles lies at the heart of the scientific activity. But internal principles alone provide only an abstract logical framework with no relevance to observation.

The bridge principles are those statements which tie the theoretical terms and the processes described in the theories' laws to the level of observations.[3] In other words, they *bridge* the gap from the theory to observations. One might go so far as to say that bridge principles are the explicit empirical *definitions* of the theoretical terms.[4] For example, they might explain how the theoretical concept of "electron" is related to such observable effects as a vapor trail in a cloud chamber. Clearly, without such bridge principles the theory is *not* testable and its essential terms are empirically meaningless.

Carl Hempel has pointed out that bridge principles need not always connect theoretical terms *directly* to observable effects.[5] They may instead relate the term to another theoretical term whose empirical meaning has already been established by bridge principles in another theory. Whether direct or indirect however, the function of bridge principles to render theories testable, is at the heart of what makes *scientific* theories different from sheer speculation.

Suppose, for example, that an extensive personal survey were made of the male students at a small California college. The survey included freshmen through seniors, and included questions on habits, personal appearance, use of time, etc. Suppose then that our pollster sat down with the data collected and attempted to find any general patterns which might account for the results. Our pollster quickly notices that the number of three-piece suits owned by any student surveyed (NS) can be easily expressed by a simple formula: NS = 3x where x is a number which ranges from 1 to 3 depending somehow on the student. Our pollster also notices that the length in inches of any surveyed student's hair (HL) can be expressed by another formulae $HL = 6 - |6-y|$, where $|6-y|$ means "the absolute value of 6−y" and y is an integer 1 to 8.[6] No one yet understands what the "x" and "y" numbers represent but it is known that the survey results can be obtained by appropriate assignment of x numbers from 1−3 and y numbers from 1−8. So far our pollster has only formulae and no theory.

$$\text{Formulae: } NS = 3x \qquad (x = 1 \ldots 3)$$

$$HL = 6 - |6-y| \qquad (y = 1, 2, 3, 4, 5, 6, 7, 8)$$

Figure 1: **Formulae**

After some careful effort our pollster announced what he called the Hair Law and the Law of Suits. The Law of Suits was that "the number of three-piece suits owned by male students in this college was three times the Worry Factor (WF) of that student. Every student had some "worries" which contributed to this factor. What a "worry" was or how to calculate the Worry Factor, our pollster could not say. The Hair Law was that hair length in inches could be found by subtracting from 6, the absolute value of the difference between the Status Number (SN) and 6. Every student had a Status Number which worked to explain his hair length, but what "status" was or how to calculate the

61

Status Number without already knowing the hair length, our pollster again could not say. For example, a student with a Worry Factor of 2/3 and Status Number of 1 was found to have 2 three-piece suits and hair 1 inch long. A student with a Worry Factor of 1 and Status Number of 8 however, had 3 three-piece suits and hair 4 inches long. These two laws are summarized as follows:

HAIR LAW: $\qquad$ $HL = 6 - |6 - SN|$ $\qquad$ $(SN = 1, 2, \ldots 8)$

LAW OF SUITS: $\qquad$ $NS = 3\ WF$ $\qquad$ $(WF = 1 \ldots 3)$

Figure 2: **Laws**

It is important to notice that the major difference between the formulae above and these laws is the introduction of the theoretical concepts, Worry Factor, and Status Number. It is not yet clear what "worries" or "status" are, neither can we measure them yet, nor do we know even whether "worries" and "status" exist, but they suggest certain causal relations yet to be elaborated; i.e., worries "cause" variations in the number of suits a student owns and status "causes" variations in hair length. Their usefulness lies in encouraging the search for causal connections and the "real" existence of worries and status.

Finally, after some abductive thinking and suddenly seeing some connections, our pollster—now turned social scientist—announces the Special Theory of Male Student Profiles. The theory shows the interconnection of four important male student characteristics. The theory consists of two laws, one additional internal principle and two bridge principles. The two laws are the Law of Suits and the Hair Law described above.

The first bridge principle is that the hitherto mysterious Status Number is the number of semesters a student has spent or that are in progress in college, as recorded by the registrar. The Status Number can be easily observed and can range from 1 for incoming freshmen, to 8 for graduating seniors. The second bridge principle ties the theoretical concept Worry Factor to observable experience. The Worry Factor is simply the number of hours each week that the student spends worrying, divided by 10. A student is worrying whenever his blood pressure is more than 15 points above normal and would report no feeling of pleasure if asked.[7] The factor can range from about 1/3 for juniors who worry very little about anything to 1/2 for seniors who spend 5 hours a week worrying about jobs for the next year up to 2 for freshmen who spend 20 hours a week worrying about leaving home, making new friends, getting dates, and passing tests.

The crucial internal principle that ties the two laws together is that the product of the Worry Factor times the Hair Length (in inches) always equals 2.[8] The Theory of Student Profile can be summarized as follows:

Laws: LAW OF SUITS: $NS = 3\,WF$ $(WF = 1, 2, 3)$

HAIR LAW: $HL = 6 - |\,6 - SN\,|$ $(SN = 1, 2, \ldots 8)$

Bridge Principles:
1. SN = Number of semesters in college as recorded by registrar
2. $WF = \dfrac{\text{Hours of worrying/week}}{10}$

Internal Principle: $WF \bullet HL = 2$

Figure 3: **Theory Summary**

    With the information given it is interesting to note that while freshmen and senior men both have conservative haircuts, more suits, and considerable anxiety (though for quite different reasons!), juniors think "suits" belong in poker and only ask "What? Me Worry?" You

might try to calculate for example hair length as a function of the Worry Factor or the number of three-piece suits owned as a function of Hair Length and to plot these results on graphs.[9]

In summary, then, we can say that theories are systems of laws and consist of theoretical terms, internal principles, and bridge principles.[10] But since laws are the parts of theories of most concern to the scientist and serve as the basis for the *use* of scientific ideas discussed in Chapter Four, let us turn to a closer look at laws.

Laws of Nature versus Accidental Generalizations

In our earlier characterization of laws we described them as generalizations containing theoretical terms. We must be more precise in defining scientific laws. In the first place, then, most laws are thought to be general statements in a *universal form.* A universal statement is a statement in the form "All X's are Y's." For example the statements, "All men are mortal," "All birds have wings," or "All Baltimore summers are hot," are universal statements.

But we must be careful because not all statements in this form should count as scientific laws. For example the statements "All bachelors are single" and "All slithy gimbies snort Nixon feathers" are in the universal form but neither is presumably the sort of statement we would want to call a scientific law. The first type of statement is called alternatively an analytic statement, a tautology, or simply a definition. The second type of statement seems to be sheer nonsense.

A second requirement in defining laws is intended to rule out such examples. *Laws must have empirical content.* In other words, laws must tell us something about the world. This quite clearly eliminates tautologies such as our example, since it tells us nothing about bachelors in the world but *only* something about how people have decided to define the word 'bachelor.' The requirement would also *seem* to rule out the nonsensical statements, but this matter is much more difficult and beyond the scope of our purposes here. It is not clear that any line can be drawn between nonsensical and meaningful statements using this criterion.[11] In *general* the application of this requirement for laws is acceptable but must remain open to modificaton in cases which are not clear cut.

An obvious third criterion for general statements to count as a law is that the statements be true. Clearly the statement "All fish fly" is a universal statement and does have empirical content in that it says something about the world. But it cannot be a scientific law simply because it is false.

But we still are not finished defining 'law.' Consider the following two true statements:

All the men at that table are Sicilians.

All unsupported bodies near the earth will fall.

While both meet our first three requirements, we would not want to say that they are both laws of nature. Only the second statement should count as a law. The first we will call an accidental generalization. The first statement does not give us ground or support for saying that "*If a man were* to sit at that table he *would* be a Sicilian.*" But the second statement does permit us to say that "*If a body were* left unsupported near the earth it *would* fall." These new statements are called 'counterfactual conditionals' because they ask "What if?" about situations which are counter-to-the present facts. Our final—and perhaps most important—requirement then for a general statement to be a 'law of nature' is that the statement must support such counterfactual conditionals. Laws of nature are presumably going to apply in all situations and times while accidental generalizations may apply to only one place or one time.

INTERPRETATIONS: WHAT DO SCIENTIFIC IDEAS *MEAN*?

Let us turn to the second major question concerning the Nature of Scientific Ideas. Do theories refer to reality and to entities which actually exist, or are they merely useful devices for understanding and even manipulating our experiences? For example, does atomic theory describe the actual microscopic behavior of matter and refer to really existing electrons, protons, and so on? Or does atomic theory merely give us a convenient way of organizing our experiences so that they seem to make sense and so that we can calculate how certain things will turn out even before we do them? There are various answers to this question both for theories and for theoretical terms. We will consider only two broad alternatives: Realism and Instrumentalism.[12]

In examining the meaning of scientific theories we are concerned most directly with the meaning of theoretical terms. We have seen above (the Distinctions section) that these terms are useful for distinguishing laws and theories from mere formulae.[13] But the purpose of theoretical terms, when tied to the level of observations, is to tell us the area of experience with which the laws of the theory deal. The term acceleration ($\overline{a}$) in the statement $\overline{F} = M\overline{a}$ tells us that the law deals with motion and not electrical phenomena. The internal principle $X = YZ$ gives us the important internal structure of this law but does not tell us what the law and the broader theory of motion really *refer to* in experience. In short, the question of what scientific theories mean is for our purposes the question of what its theoretical terms mean.

The Realist View of Theories
The realist view takes theoretical terms as referring to physical

entities (or their properties) which really exist. In general these entities may be of two sorts. 1) They may be directly observable and tied to the theoretical term by an explicit definition. Such explicit definitions are provided by the theory's bridge principles and give the theoretical term *empirical* meaning. 2) The referent entities of a theoretical term may not be directly observable and are defined only implicitly by the context in which the theoretical term occurs in the theory. Such implicit definitions are provided largely by the theory's internal principles and give the theoretical term *contextual* meaning.[14] What is important however is that for the realist, theoretical terms really do refer to something outside of the theory regardless of how they are defined. The terms 'atom,' 'gene,' 'force,' and 'electron' all refer to actually existing entities or their properties.

The realist's view of reference implies first that theories in science can be true or false.[15] A theory of light which includes a theoretical term 'ether' will be either true or false in its reference to reality. Second, theoretical terms generally should be translatable into statements using observation terms only, since presumably these terms are referring to real physical entities or their properties. Each of these implications leads to certain problems.

The first problem facing the realist view of theories and theoretical terms is that it does not seem to account for the fact that apparently incompatible theories are sometimes used for the same subject matter.[16] Liquids are sometimes treated as systems of discrete particles and sometimes as continuous media. Light seems to be a wave on Mondays, Wednesdays, and Fridays but a system of particles on Tuesdays, Thursdays and Saturdays. How can both theories with their theoretical terms be *true* about the actually existing entity? A possible reply is that where such incompatibilities exist we simply do not *yet* have an adequate theory.

A second problem is to explain how the realist can ever hope to provide translations of theoretical terms into observational terms when the theoretical terms *may* be only implicitly defined by context and its referent entities at best only indirectly observable. This is a serious problem for the realist. How could 'electron' ever be translated into a set of observation statements? Even if an infinite set of such statements were permitted, remaining thus always incomplete, the fact that the term's meaning is largely contextual makes its translation virtually impossible.

Another version of the realist view is sometimes called the descriptivist view. This view arose with the idea that science does not explain anything but merely describes things. It has been very influential in the philosophy of science especially in the late 19th century as scientists became uneasy about making their commitments to the real reference of theories too strong. The descriptivist view permitted theoretical terms only if they were *explicitly* definable by a finite set of observation statements. Theoretical terms then became merely convenient shorthand summaries of these observation statements.

The descriptivist view might seem to avoid the problem of translating theoretical terms which are only *implicitly* defined, i.e., because it does not accept such theoretical terms. Nevertheless, the problems here are just as serious. In the first place those who tried the translation task of even the simplest theoretical terms found it impossible because there was no adequate observation language to use. The search for such a language was abandoned. A second problem—not unrelated to the first—is that the fundamental distinction of theoretical versus observation terms has itself been disputed and even generally refuted.[17] This meant that observation terms and theoretical terms could never be completely distinguished. Translations of theoretical terms would be impossible both for the descriptive view and the realist view. This problem virtually eliminates the descriptivist position and forces the realist to rely far more heavily on the implicit definitions to give contextual meaning to theoretical terms. The problem which remains is to do this without sacrificing the fundamental realist assumption that such terms do *refer* to actually existing entities or properties.

The Instrumentalist View of Theories
The instrumentalist view assumes that theoretical terms do not refer to actually existing entities or properties. Such terms are instead, merely convenient tools or devices for organizing our experiences. They involve no metaphysical commitment; i.e., they do not commit the scientist to physical entities beyond the terms. E. Nagel has used the illustration of a hammer to clarify the instrumentalist position.[18]

> A hammer is a deliberately contrived tool, with the help of which a variety of "raw materials" can be brought into definite relations, so as to yield such things as packing boxes, furniture and buildings. . . . We would think it nonsense were anyone to suggest that a hammer is in any familiar sense "equivalent" to the things produced or producible by its means; and we would also regard as curious the question whether a hammer adequately "represents" the products already made. . . .

For the instrumentalist theories are like hammers and other tools in important ways. Theories are simply tools or instruments of the mind, used to direct our scientific experimentation and to show connections between areas of experience that might otherwise appear unrelated.

They make our experiences (raw materials) useful (packing crates). But theories do not represent (realist) or even summarize (descriptivist) our experience. They do *refer* to experience but only to organize it for use.

The instrumentalist view that theories do not have factual reference has important implications. In the first place it implies that theoretical terms cannot be translated into observation terms. They *have* no referents in the physical world. Second, from the first implication it follows that theories will not be either true or false in the "correspondence" to the world sense since they don't refer to the world at all. Theories will either be useful or not useful. Incompatible theories of the same subject matter (*e.g.*, wave or particle theories of light) present no difficulty for the instrumentalist. They can *both* be useful. While these implications seem to avoid some of the realist's problems, the instrumentalist view has its own problems.

One problem for the instrumentalist is that he must concede that the entities described by their theoretical terms do not really exist. But this means denying reality to atoms, molecules, electrons, and so on. Instrumentalists often contradict themselves and insist that these entities are real. A second problem is that experiments used to confirm theories seem to be a waste of time if theories can only be useful and never true. Finally, the instrumentalist must face the almost inevitable relativism of his view. If theories are never true then progress in science must be defined in terms of usefulness. But usefulness is at least in part relative to the historical context. Most practicing scientists are committed to a more objective view of their activity and progress.

This discussion of what theories mean has been left open-ended. We have not treated all possible views nor have we stated all the defense an instrumentalist or realist might give to the objections we have raised. It is sufficient to have drawn the basic distinction and to have suggested that neither view is wholly without difficulty. While contemporary physics may tend to be interpreted instrumentally, it remains true that most practicing scientists are realistically inclined.[19]

SUMMARY

In this chapter we have examined the nature of scientific ideas. We have seen that the activity of science produces laws and theories whose structure and meaning can be examined. Such examination gives us a perspective of science as something fixed or static rather than as a process. In the second and third sections we investigated the make-up of laws and theories, looking at their component parts: theoretical terms, internal and bridge principles. In the fourth section, we turned to ask about what theories mean and discussed two basic alternatives, each with its own set of problems. Having investigated how scientific ideas are formed, how they develop, and in this chapter what they are and mean, there remains what may be the most interesting question of all. How are scientific.ideas used? This is the subject of Chapter Four.

CHAPTER FOUR
The Use of
Scientific Ideas:
Explanation

INTRODUCTION

When scientific ideas have been fairly well confirmed, what is done with them? Often they are written down in textbooks as collections of facts or knowledge. These collections constitute science as something you *have,* i.e., as a noun, rather than as something you *do.* Looking at these "finished products" as they are collected in textbooks gives us the "static" perspective of science.[1] In Chapter Three we examined those "finished products" to understand their nature, i.e., what they are and what they mean. But to what end are these products collected there? What is the *use* of scientific ideas? This is the central question for our discussion.

We suggested in Chapter One that tentative theories are invented for the purpose of *answering* or *accounting* for a "problem."[2] It was also suggested there that the purpose of science was to produce "conceptual generalizations *about* the many particulars of sense experience." The meaning of each of the italicized words in these statements can be summed up in the concept of "explanation." The primary use of scientific theories is to *explain* problems, or put another way, to *explain* the many particulars of sense experience. We will see that explanations may take many forms but that the "covering-law model" has dominated for many years. The *main point* of the chapter and thus the answer to its central question—is that scientific ideas are used to *explain experience* generally according to a covering-law model.

TYPES OF EXPLANATION

Illustrations
What in general does it mean to *explain* something? Let us look at several illustrations of the explaining process.

1.   For example, what does it mean to explain the fact that a certain dark plum fell to the floor when released from my hand several

years ago. The fact to be explained here is the occurrence of a specific individual event. In broad terms the explanation might be that when it was released, the plum was unsupported in mid-air and that generally speaking, gravity makes unsupported bodies fall.

2.   How do we explain the fact that Henry VIII of England sought to annul his marriage to Catherine of Aragon? Usually it is said that he did so because he wanted a male heir to his throne. Catherine could bear him no sons and he wished to remarry. He could not do so because he was already married to Catherine and, of course, divorce was prohibited by the Church.

3.   Why do human beings have a heart? One might explain this by saying that the body needs to have blood pumped to all parts of the body and the heart serves this purpose or function.

4.   How do you explain the structure of the atom? One might answer by saying that, "The atom is like a miniature solar system with a nucleus like the sun and electrons in orbits like the planets." Features of those orbits might be best explained by likening them to lanes on a wide expressway.[3]

From each of these otherwise quite different illustrations of explanation one common feature emerges. Each example of explanation involves two essential—perhaps obvious—elements. First, each involves something-to-be-explained. An answer without a question is quite useless. To avoid the awkward hyphenated phrase, the word 'explanandum' is used to refer to the thing-to-be-explained. Second, every explanation involves the thing-doing-the-explaining, abbreviated as the 'explanans.'[4] The explanans is the answer to the question.

But while all explanations involve both an explanandum and explanans, they do not all exhibit the same kind of relationship between the two. The differences in this relationship distinguish the various types of explanation, which help us in turn to classify examples of explanation such as those given above. Let us examine three such *types* of explanation.[5]

Teleological Explanation

At the time of Aristotle, and even before, to explain something, usually meant to answer the question *Why*. To explain the growth of acorns, the fall of rain, or the edicts of a king meant to state *why* acorns grew, rain fell, and kings commanded. The answers provided were in the form of purposes or goals. (*Teleos* is the Greek word for purpose or goal.) Acorns grew *in order to become oak trees*. To become oak trees was their goal and this goal thus explained their behavior. The explanandum is the question, "Why do acorns grow?" The explanans is the purpose expressed in the italicized phrase *in order to become oak trees.* Likewise, rain falls because all bodies have as their goal, *to be at the center of the earth.*

So far you may laugh and say that this first type of explanation really explains nothing at all and is obviously obsolete. But to answer the question, "Why does the king command?" (explanandum) with the statement, "His purpose is to keep peace in his land" (explanans), seems to be a perfectly acceptable explanation. Furthermore, in the second illustration above it makes good sense to explain Henry VIII's behavior by pointing out his desire (goal) to have a male heir. And in the third illustration, it is reasonable to answer the question, "Why do humans have hearts?" by saying, "It is the function (purpose) of the heart to pump blood to body cells." In fact, this last example illustrates how teleological explanation has even been called 'biological' or 'functional' explanation, though this type of explanation is far less common in our age of biochemistry.

The obvious point is that today teleological explanations are sometimes quite acceptable while at other times such explanations appear ridiculous. Of course, for Aristotle, answering *why* by stating a purpose always made sense.

By at least the 17th century and the time of Newton, it was generally accepted that explanations which attributed purposes to everything in the world were not adequate. Could acorns and raindrops really have "purposes?" Was the world really like that? Answering the question *why* in such terms became increasingly dubious.

A new approach to explanation began to emerge. This new approach was that, in science at least, the question to be answered by an explanation would be the question *how* and not the question *why*. Explaining became more a matter of *description* than anything else.[6] This important shift was promoted by the fact that during this same period, there occurred a tremendous advance in the field of mathematics. The birth of calculus as a powerful descriptive tool facilitated the formation of general descriptions for complex phenomena. These general descriptions are precisely the scientific theories discussed above. This shift in explanation type with the simultaneous advance in mathematics so spurred the progress of science that this period is said to mark the birth of modern science.

But to say only that explanation was now a matter of answering the question *how* does not yet specify the *way* that question was (and is) answered. Let us consider two possibilities: analogical, and scientific explanation, each of which answers *how* in a different way.

Analogical Explanation

Some have suggested that to explain something is to *compare it by analogy to something familiar.* The explanans is said to describe the explanandum as a model describes the thing it copies. A good characterization of this type of explanation reveals three important requirements:

1. The explanandum is something complex or unfamiliar.
2. The explanans is relatively familiar and
3. The relationship between the two is one of analogy.

For example, radio waves, which are relatively complex and unfamiliar might be explained by stating that *they are like ripples on the surface of a pond.* Notice that the explanans (in italics) is both familiar to us and related to the explanandum (radio waves) by the analogy implied in the word 'like.' The fourth illustration given above is another example of analogical explanation. The explanandum is the structure of the atom. The explanans is the relatively more familiar solar system. The relationship once again is one of similarity or analogy.

Let us consider a final illustration of this kind of explanation at work. Suppose in a review session, the night before an exam, you asked me to explain the process of genetic transcription in a living cell. An analogical explanation would describe how that process works but *not* state why it happens. I might begin by saying that it is a process of decoding biological information stored sequentially on polymers. For some of you, that explanation would perhaps suffice, because you are familiar with the nature of biological information and with polymers. Most of you however would probably snort scornfully: "Riiiiight! That explains a lot!" Because I want to please, I might go on then to say that the process is very much like the playback operation of a magnetic tape head. Perhaps another group of you would sit back, relax, and say, "Oh, I see," because you are familiar with the tape recording operation. By this time, however, many of you would be quite irritated by all this flaunting of scientific trivia, though perhaps a little impressed by those people who had already sat back relaxed, (C. P. Snow rides again) and maybe even a little panicky thinking about tomorrow. Seeing the look on your face, I might go on, "Well, it is really a lot like a carpenter using a template." Another group of you see the light . . . the rest of you are storming out the door. In desperation I ask, "Have you ever copied a key in clay? It's a little like that!" For most of you, that experience is familiar enough that you begin to understand.

The key to understanding analogical explanation is that such explanations must by definition reduce the unfamiliar to the familiar. Hence *such explanations are always psychologically satisfying.*

Until the questioner is satisfied, the explanation is incomplete. When similarity has been established between the explanandum and explanans we feel that we understand the explanandum; it has been explained.

At least two major problems arise immediately with this type of explanation. First, what counts as an explanation for one person is *not* an acceptable explanation for another. What is acceptable (i.e., psychologically satisfying) is relative to the varying backgrounds of the questioner. This fact is obvious from the review session illustration. Some students with biology backgrounds understood one explanation. Those with some physics were satisfied with another. The students with knowledge of carpentry accepted a third, and the rest understood the fourth because of their auto theft experience. This "relativity to background" makes analogical explanations often ambiguous.

Second, and most important, there is some question whether analogical explanations really *explain* anything at all! It is common knowledge that metaphors are ambiguous and that analogies "limp" because they are always incomplete. Yet these are precisely the devices used to do the explaining in this type of explanation. While analogies may be useful to illustrate a point, do they really explain anything? They certainly give us a sense of psychological satisfaction or relief, but is that enough?[7]

Scientific Explanation

For any explanation to be a scientific one, it must meet certain requirements in answering the question *how*.

First, any scientific explanation must be *relevant* to the thing needing explanation. In other words the explanans must be relevant to the explanandum. 'Relevance' is hard to define. In general, it means that there must be some kind of connection apparent between the question and the answer. Now that may seem an obvious requirement to you, but its importance has not always been stressed, even by those who have called themselves scientists.

Take, for example the astronomer Francesco Sizi, who offered the following "explanation" why there could not possibly be satellites circling the planet Jupiter, although Galileo claimed to have already observed them directly through his telescope.

> There are seven windows in the head, two nostrils, two ears, two eyes and a mouth; so in the heavens there are two favorable stars, two unpropitious, two luminaries, and Mercury alone undecided and indifferent. From which, and many other similar phenomena of nature such as the seven metals, etc., which it were tedious to enumerate, we gather that the number of planets is necessarily seven. . . .

Not content with his refutation yet, Sizi goes on to a final crescendo of stunning logic:

> Moreover, the satellites are invisible to the naked eye, and therefore can have no influence on the earth and therefore would be useless and therefore do not exist![8]

It would appear that Francesco Sizi did not want to be confused with the facts.

From this example it is obvious that analogies between explanandum and explanans are *not* always relevant. It is also apparent that what may sound like logic need not carry its force of thought. The example does not tell us so much what relevance is as what it is not. But any model of explanation which tries to fill this requirement must surely lead us "naturally" from the answer to the question. It might just give us the best grounds possible for acting in regard of the explanandum. The explanans might lead us to expect the explanandum. It might make the explanandum unsurprising. And perhaps the explanans may be so relevant as to make the explanandum even predictable under certain circumstances.

A second requirement for any scientific explanation is that it be *testable*. This requirement is not surprising, because as suggested in Chapter One, the ultimate criterion of truth in science is sense experience. If that criterion applies to science as an activity, it is to be expected that it will likewise apply to the use of those theories developed by that activity. When a theory is offered as the scientific explanation for an event, this requirement is that the connection between theory and event be supportable by sense experience.

Suppose, by way of a counter-example, I proposed that the phenomenon of tides (explanandum) could be explained in the following way (explanans): In the Pacific Ocean, between Hawaii and Pago Pago, there lives a giant dragon by the name of Milford. He is as large as Alaska and Rhode Island combined. Because he is very old, Milford weighs nothing at all, and takes up no space. His original chartreuse and violet colors have long since faded, so that now he is utterly invisible. Living in the water, he keeps himself quite clean—even brushing thrice daily with Pearl Drops—and does not smell. Because he once taught in the Schools of Fishes, he lost his voice and was forced into early retirement at the age of only 2,300 years. In short, Milford is your typical, colorless, untastable (it's hard to call such a guy "tasteless"), untouchable, quiet, dragon. But Milford is very friendly, and as the moon rises and sets daily he greets it with a wag of his incredibly large tail. (You see Milford's eyes are failing too and he mistakes it for a smiling face.) The motion of his tail naturally causes the ocean water to slosh back and forth in its bed and this in turn affects the other oceans throughout the world. The net effect is what we call the tides.

While my explanation is perfectly relevant to the explanandum at hand (tides), it is utterly untestable and hence thoroughly unscientific. Notice that the same would probably be true for any explanation involving God or human purposes, desires, and emotions.

It is possible to judge (as we will do below) whether a given *type* of explanation meets these first two scientific requirements. But there are three other requirements which must also be fulfilled that can be judged only by taking each particular explanation individually. While these will not be of further concern to us here, it is important to mention them. A scientific explanation must 1) be compatible with other existing explanations, 2) show potential explanatory scope (i.e., be able to explain many new things), and 3) be fundamentally simple in concept. The last two are comparative requirements in that they can be used to compare and decide between alternative scientific explanations which are equal in every other way.

In this section of the chapter we have looked at explanation in general. This is so because our central question is "How are Scientific Theories Used?" The answer from this section is that they are used to *explain* experience. We have seen further that explanations always consist of a question or thing-to-be-explained (explanandum) and an answer or thing-doing-the-explaining (explanans). But, of course, there can be different types of explanations that can be used, even to explain the same explanandum. *Teleological explanations* answer the question *why* about their explananda by citing purposes or goals as explanans. *Analogical explanations* answer the question *how* about their explananda by citing familiar analogies as explanans. *Scientific explanations* are the subject of the section which follows, but must in general be both relevant and testable.

SCIENTIFIC EXPLANATION

The main point of this chapter is that "Scientific theories are used to *explain* experience generally according to a *covering-law model.*" We have focused on the first italicized concept in the previous sections. In this section we will consider the particular nature of the covering-law type of scientific explanation.

Recently, in modern science, theories have been *used* to explain experience most frequently according to a very precisely defined pattern or model. This type of explanation meets the general requirements for scientific explanations that we discussed in the last section. There are various names for this model. But because it was best—if not originally—set forth by C. Hempel and P. Oppenheim, the model is frequently referred to as the 'Hempel-Oppenheim' model of scientific explanation.[9] Because it involves the use of both deduction and scientific laws (from the Greek *nomos*) it is also referred to as the 'Deductive-Nomological' model or just the 'D-N' model for short. Finally because of the way the laws are used to explain, it is also frequently called the 'Covering-Law' model of explanation. These names are synonymous for our purposes.

Each branch of science has had a different degree of success in forming well-confirmed theories. Because these theories are required by the covering-law model, each branch is able to use that model to varying degrees. In physics or chemistry, for example, the strength of current theories makes use of this model the standard. But in the social sciences, the lack of consensus and confirmation on most existing social theories makes use of the covering-law model nearly impossible. The importance of using the strictly scientific pattern for explanation is underscored by the fact that some have suggested that unless a discipline can use this model it cannot legitimately claim the status of a science. The covering-law model, on this view, defines science itself. While this may be extreme, it is at least true that the use of covering-law explanations is highly desirable and seems to be a goal for many branches of science.

Components

Covering-law explanations, like all others, involve an explanandum, an explanans, and a distinguishing relationship between the two.[10]

The explanandum, as usual, is some *particular* event or phenomenon which needs to be explained. It may be a specific event, like the 1977 New York blackout, or a class of events like tides or billiard ball collisions. It is usually thought to need explanation because it is unexpected or surprising to the person requiring the explanation. It is also normally an event or phenomenon in sense experience (which either directly or indirectly affects sense experience).

The explanans always consists chiefly of scientific laws or theories. From what we have learned, we know that such laws or theories are the products of the scientific activity, and once fairly well confirmed, are often found printed in texts. These theories we also know to be conceptual generalizations. As such, we have seen that they are sophisticated descriptions of nature.[11]

But the key component of the covering-law model is the distinctive relationship which must obtain between the explanandum and explanans. As in any explanation type, it is this relationship that distinguishes the covering-law model from others. In this case, it is a *deductive* relationship. By this is meant that *the explanandum follows by the logic of deduction from the explanans.* This statement serves admirably as a definition for the covering-law model of explanation. When this relationship holds between the explanans and explanandum we say that the explanans "covers" or "explains" the explanandum.

Examples

For example, you might, using this model, explain the fact that "Tuesday's lecture was boring" (explanandum) by *using* the general (and well confirmed) theory that "All lectures are boring" (explanans). Notice of course that the explanandum follows by the rules of deduction from the explanans. We say that the theory "covers" the event needing explanation.

Take another example. Suppose you asked me to explain scientifically the fact that the apple you were just eating, just fell on the floor (explanandum). Using the covering-law model I might respond something as follows: "Well, first the Law of Earth's Gravity says that 'All unsupported bodies fall,' and second, your apple was an unsupported body." These two statements together constitute the explanans and are said to "cover" or explain the explanandum. This explanation can be diagrammed for clarification:

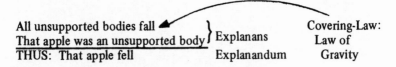

All unsupported bodies fall	⎫ Explanans	Covering-Law:
That apple was an unsupported body	⎭	Law of
THUS: That apple fell	Explanandum	Gravity

Figure 1: **Covering-Law Explanation of a Falling Apple**

Notice the two-part explanans and the deductive relation between explanans and explanandum as indicated by the connecting word

thus, and by the heavy dark line signalling the conclusion of a deduction. Finally notice that the covering-law in this case is the well confirmed Law of Gravity.[12] The word *cover* is used, because the *general* law can be seen to account for (cover) an infinite number of such *particular* events. The explanandum is always just one particular instance of the general law which covers it. We can think of the law as an umbrella which can cover many different explananda.

Covering Relation
To further illuminate this essential concept of covering let us review our understanding of deduction. In Chapter One we observed that deduction proceeds according to strict rules and is almost mechanical or automatic. We also observed that it carries a force of necessity. This force must then also apply to covering-law explanations because of course the covering relation between explanans and explanandum is deductive.

A deductive relation between two things, A and B, can be abbreviated by an arrow: A→B. A and B stand for any two statements that stand in this relation; *e.g.,* "All lectures are boring" (A), and, "This lecture is boring" (B), or "It is raining" (A), and, "The streets are wet" (B).

There is a point to be made about A→B which is important for us. We can say that A is a *sufficient condition* for B. In other words, if A is true then that is sufficient information to guarantee that B is also true. Another way of saying what is really the same thing is that B is a *necessary condition* of A. In other words, B is necessarily true if A is true.

These two statements accordingly apply to covering-law explanations and help us finally to see why this model of explanation is so desirable in science. We know from our definition that the explanandum follows by deduction from the explanans. This means that if the explanandum is B and the explanans is A then our covering-law explanation can be represented as A→B. But then we also know that the explanans is a *sufficient condition* for the explanandum and the explanandum is a *necessary condition* of the explanans.

Column I		Column II	Column Relationship:
A	→	B	I sufficient for II,
Explanans	→	Explanandum	II necessary of I
Example:			
"All lectures are boring"	→	"This lecture is boring"	

Figure 2: **D-N Explanations**

Since we know that it is the explanans which covers the explanandum we conclude that the phrase "to cover" means "to give sufficient condition for." This is the strength of scientific explanation; it uses theories *to give us sufficient conditions for expecting* the things we are trying to explain. When the explanandum is then expected and no longer surprising, it is explained.

Covering-Law Variations

1.   For the purpose of clarity, the examples chosen to illustrate the application of the covering-law model have been very simple. Yet it borders on the ridiculous to explain boring lectures, or falling apples, simply by saying *all* lectures are boring, and *all* unsupported bodies fall. In reality, applications of this explanatory pattern will involve explanans which include many many premises, laws, and theories, just as the apple example contained two. Often some of these premises are so obvious they are not stated explicitly, but are suppressed. We call such explanations, 'elliptical' or 'enthymematic' covering-law explanations.

2.   The covering-laws we have used in our examples have

been generalizations of the *universal* sort, i.e., they cover *all* cases. "*All* lectures are boring" and "*All* unsupported bodies fall."[13] Rarely, however, do scientific laws or theories take this form. Often they are statistical in nature.

When such statistical laws are used to explain statistical events, the explanation *can remain strictly deductive*. For example, the fact that six million American cigarette smokers got lung cancer last year can be strictly covered by the explanans consisting of the statistical law that "6% of all American cigarette smokers get lung cancer," and the additional premise that there are 100 million Americans who smoke cigarettes.

When, however, such statistical laws are used to explain *particular* events, the explanation is *not strictly deductive* and hence is not technically a covering-law explanation at all. Instead of being called a Deductive-Nomological (D-N) explanation which would be a *deductive* argument based on *laws* (*Nomos*), such an explanation is called Inductive-Statistical (I-S) because it is an *inductive* (i.e., statistical or probabilistic) argument based on statistical generalizations. For example, the fact that "Sam had a heart attack," can*not* strictly be covered by the explanans consisting of the statistical law "90% of the people in profession X have heart attacks," and, "Sam is in profession X." The explanans is obviously *not* a sufficient condition for the explanandum.[14] This explanation may be represented as follows:

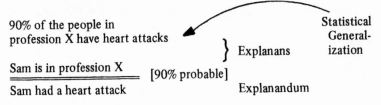

Figure 3: I-S Explanations

The double line and bracketed number mean that the explanatory argument is not deductive but probabilistic (inductive) and has a 90% chance of being valid. Such I-S explanations are usually acceptable only when the probability is 50% or more, i.e., when the explanans really does lead us to *expect* the explanandum.[15]

Of course, in a sense, statistical or probabilistic explanation approaches the covering-law's deductive ideal, so there may be question as to whether it is really a distinct type of explanation all its own. It is important to see that explanations are not statistical whenever statistical premises are involved, but only when the connection between explanans and explanandum is probabilistic (inductive) instead of deductive.[16]

The Explanation-Prediction Similarity

It should already be apparent that there is a great deal of similarity between covering-law explanation and the process of prediction discussed in Chapter One. The chief similarity is that they both involve a deductive process or relationship. In both cases, something general leads by deduction to something particular. In prediction, the tentative theory leads by deduction to the experiments to be performed in testing the theory. In explanation, that theory has now been well confirmed and serves as the explanans which leads by deduction to the explanandum.

From this we can see that covering-law explanation is in fact just a matter of *predicting* the thing-to-be-explained. To predict is the same as to "give sufficient condition to expect." The logic is identical.

One obvious difference, however, is that of *purpose*. While prediction is part of the process of *forming* the scientific theory (especially, testing it), explanation is the process of *using* the scientific theory. In prediction the theory is tentative, in explanation the theory (explanans) is well-confirmed.

Another difference is really only one of *perspective*. In both cases, the *logic* (deduction) goes from general to particular (Theory to Experiment and Explanans to Explanandum). However, in *forming* theories, the process proceeds from tentative theory (general) to experiment (particular) while in *using* these theories for explanation the process begins (naturally) with the question or explanandum (particular) and ends with the answer or explanans (general). No one would try an experiment before he knew the theory he was testing. Nor would anyone provide an answer without knowing the question. So while Prediction and Explanation are logically identical, they differ in purpose and perspective.[17]

	Logic	Purpose	Perspective
Prediction	Deductive (General→Particular)	Forming Theories	Theory *then* Experiment (General)　　(Particular)
Explanation	Deductive (General→Particular)	Using Theories	Question *then* Answer (Particular)　　(General)

Figure 4: Prediction-Explanation Similarity

84

Meeting the Requirements of Relevance and Testability

Having carefully examined the covering-law model of explanation, it remains to ask whether it does indeed fulfill the requirements for a scientific explanation which were established in the previous section.

In discussing the first requirement, that of relevance, no precise definition could be given. Many characterizations were offered. In general, we said that a relevant explanation would show "some kind of connection between the question and answer." More specifically we suggested that relevant explanations "might lead us naturally from the answer to the question," "might make us expect the explanandum," and might "make the explanandum unsurprising." While Analogical explanations may meet this requirement in part, the covering-law model obviously does so in a much stronger sense. The "connection" it shows is that of deduction. The explanans leads "naturally," in fact, necessarily, to the explanandum. By giving us "sufficient conditions to expect" the explanandum, i.e., predicting it, the explanans makes the explanandum *completely* unsurprising. In short, the covering-law pattern of explanation meets the requirement of relevance in the strongest possible way.

The model is equally successful in meeting the requirement of testability. First the covering-law which serves as the explanans in this model is itself a scientific law or theory which has been formed through a process which we have already seen to involve testing against the standard of experience. These theories are not used for explanation until they have been well confirmed (i.e., tested). Second, theories are tested by their empirical predictions (implications). But we can be sure that the theories used as explanans in covering-law explanations have at least *some* empirical implications, and thus are testable because the explanandum *itself* is one of these. In short, the covering-law pattern of explanation meets this requirement because of the testing involved in forming the explanans it uses and for the same reason it met the first, namely, its similarity to prediction.

Counterexamples to the D-N/I-S Models

But the explanation-prediction symmetry which seems to make the covering-law model of explanation so attractive is also responsible for some of its difficulties. The symmetry seems to imply that good predictions are always good explanations and vice versa. Meeting the D-N requirements for explanation would appear to be sufficient for giving a good explanation and if something is a good explanation it must necessarily meet the D-N requirements. But there seem to be counterexamples to both the sufficiency and necessity of the D-N requirements as conditions for good explanations.

85

1.   Sufficiency – Imagine a flagpole standing in the open sun on level ground. Suppose it casts a shadow which extends to a point 50 feet away and suppose that the angle at that point between the ground and a light ray from the top of the pole is 45°. Now any high school trigonometry student could deduce from these conditions and the laws of optics that the flagpole is 50 feet high. This deduction would satisfy the D-N requirements for an *explanation* of the fact that the flagpole is 50 feet high. And it does allow us to predict the flagpole's height, testable later by a climbing experiment involving a tape measure. But no one would want to accept the shadow and its measurement as a good *explanation* for the flagpole's height. The D-N model with its predictive power is apparently not sufficient for a good explanation.[18] The Babylonians were excellent predictors of astronomical events but they had no scientific understanding or explanation of astronomy at all.[19]

2.   Necessity – But while a D-N explanation and its symmetric predictive power may not be sufficient to give us a good explanation, surely a good explanation, whatever else it involves, will necessarily also involve the D-N requirements and particularly the predictive power which makes the explanans expected and unsurprising. Surely any good explanation will do this.

Consider however the following counterexample.[20] It is a known medical fact that a small percentage (let us say 2% for discussion) of those persons injected with penicillin develop a rash. Suppose Jones develops a rash. What is a good explanation? It is, of course, that he was injected with penicillin. But if this explanation were put in the covering-law form it would not satisfy the I-S requirements because it does *not* give us good reason to expect, i.e., predict Jones' rash.

2% of those injected with Penicillin develop a rash	
Jones was injected with penicillin [2%]	} Explanans
Jones got a rash	Explanandum

Figure 5: Unacceptable I-S Explanation

The bracketed probability does not give us good evidence to expect the conclusion. So it appears that what we would intuitively want to accept as a good explanation of Jones' rash would not *necessarily* meet the covering-law requirement that explanations make the explanandum expected or predicted.

These counterexamples alone do not refute the covering-law model. Advocates of the model continue to defend it in a variety of ways.[21] Nevertheless, the examples create sufficient doubt to warrant mention here of at least one alternative model of explanation; the statistical-relevance (S-R) model.

The Statistical-Relevance Alternative

It appears that predictability as part of explanation is neither sufficient nor necessary for a good explanation. This predictive power is what allowed the covering-law model to meet the criterion of relevance for scientific explanations. Apparently predictability meets the relevance requirement *too* strongly; in fact in the strongest possible way. But if predictability is too strong, how else can an explanation be relevant?

W. Salmon has suggested what he calls the Statistical-Relevance model for explanation.[22] While it is too complex to discuss in detail, it warrants a brief description. Two features distinguish it from Hempel's Inductive-Statistical (I-S) model. In the first place, the explanation is *not* an argument of any sort, either deductive like Hempel's D-N model or inductive (probabilistic) like Hempel's I-S model. For Salmon, explanations consist of simply stating a list of probability laws, only one of which applies to the explanandum. The second difference follows from the first. Because the probability laws listed may involve high or low probabilities, the one which applies to the explanandum may make the explanandum likely or unlikely. In other words, unlike Hempel's model, the explanation may fit the S-R model without making the explanandum "expected" or "predicted." The explanations do always give the best *possible* grounds for expecting the explanandum but those grounds may make the explanandum quite *unlikely* and still serve as a good explanation. The relevance of such an explanation is tied up in the fact that the probability laws are the *most relevant* laws possible. Any further discussion of this feature is beyond the scope of our discussion.[23]

SUMMARY

In this important chapter we have seen that scientific ideas are *used* to explain experience. We have investigated the general nature of explanation and examined two possible models of explanation which fail to meet the requirements of relevance and testability for scientific explanations. We have discussed the covering-law model in both its D-N and I-S form and shown that it predominates as the chief model for scientific explanation. Finally, however, we have observed that the covering-law model is not without counterexamples which forced us to consider another alternative, the Statistical-Relevance model.

In conclusion, we see that the question of what counts as a

good scientific explanation has not been fully answered. To observe this fact is to be humble about the primary purpose and use of science. While this limitation *may* only be a limitation of philosophers who try to clarify what the scientists are doing, there are serious limitations to science itself to which we must now turn.

CHAPTER FIVE
The Limitations of Scientific Ideas

INTRODUCTION

In Chapters One and Two we examined the *formation* and *development* of scientific ideas respectively. This provided a perspective of science as an activity. Scientific ideas are formed by a process which is both creative and cyclic. They develop through history in a way which is both evolutionary and revolutionary. In Chapters Three and Four we turned to look at science as a finished product. We saw something of the *nature* and *use* of scientific ideas. They are abstract and general but their meaning is open to alternative interpretations. Once well-confirmed, these ideas are used by scientist and layman alike to explain events in experience, chiefly according to a "covering-law" model of explanation.

In this chapter we will focus on a very natural question: How certain can we be of these theories which science produces and uses? Are they trustworthy? In short, we are concerned with the limitations of science. Of course this question cuts two ways. We must assess not only the degree of uncertainty in science, but must also examine its strengths. The *main point* of what follows is to provide a healthy balance of appreciation and scepticism regarding the claims of science.

Historically, science has had both periods of greater and of lesser acceptance. At the time of Plato, it was not popular to be concerned with sense experience and the world around us. That sensible world was only an imperfect shadowy copy of the "really real" world of what Plato called "Forms." The purpose of life was to be concerned to know these Forms and especially the Form of the Good. So it was foolishness to be a scientist.

But Plato's student, Aristotle, changed all that and rehabilitated the worth of nature. His careful studies led him to observe many things about the habits of animals and fish, about the features of plants, and the patterns of the heavens. For him the world of sense experience was "really real." One might even say that Aristotle was among the earliest true scientists, investigating nature for its own sake.

Unfortunately, Aristotle's contributions were largely forgotten. Science fell into relative disrepute for over a thousand years up to and including most of the Middle Ages. During this time, the Church exerted by far the most significant influence. It was often felt that scientific theories were unnecessary in answering questions about the world. Scriptures provided all that was important, especially in answering the *why* type of question.

Beginning in the 13th century with men such as St. Thomas Aquinas, there was a re-discovery of Aristotle and renewed attention to man's reason and to the world of sense experience. Science received an especially important boost during the Renaissance and in the 17th century with the work of Bacon and Kepler, Galileo and Newton. The advance in mathematics promoted this rapid development. Modern science was born.

In general, science has grown steadily in reputation since that time. Because it led, indirectly through technology, to the tremendous material progress of the Industrial Revolution and to so many of the obvious features of society today, science has been thought to have an answer for all of man's problems. Its limitations have gone uninvestigated to the point where science has become like a religion itself, to be worshipped, with white-robed "priests" and all. Using the word 'scientific' seems to carry with it a magic of truth, objectivity, and absolute certainty. From soap powders to dog food, 'scientific' always means "better." Especially in the 1950's and 1960's no manufacturer would dare introduce a product without one or more scientific studies to back it up. The danger, of course, occurs when the scientist-priests of this new religion themselves begin to believe in its limitless infallibility. If Chapter One above is directed to the layman, then much of this chapter is directed to the scientific side of C. P. Snow's "gap."

It must be mentioned that perhaps some of the shine has worn off of science's image in the 1970's. The priest's white robes have been smudged. This disillusionment has perhaps arisen from science's failure to solve many of the pressing social problems of our age, from poverty to world peace. In fact, it appears that science may have created as many new problems as it has solved old ones, often plunging ahead without adequately investigating the impact of its action in other areas. The rapid increase in energy demand, prompted not only a serious energy problem but a pollution problem as well. Far from bringing peace, science seems to raise the stakes in an ever more costly and uneasy armament balance. Even in medicine and agriculture, the cure is sometimes worse than the problem. Science, in short, is perhaps not worshipped, as it was only 20 years ago.

The point is to illustrate that science has had its "ups and downs" in reputation. We have tried to get behind the popular current view of science—whatever that may be—and to investigate those factors which give science both its certainty and uncertainty.

SCIENCE'S CERTAINTY

In discussing the certainty of science we may focus our inquiry perhaps by asking specifically, "What advantage does scientific knowledge have over other kinds?" As with nearly every question, we can answer this one on different levels of complexity.

On a fairly crude level we can respond by saying that science has an advantage in being concerned only to answer limited kinds of questions about the world. It is able to be successful in the questions it answers because it does not attempt to answer all possible questions about all things. We have already noted how science seemed to spurt ahead in the 17th century when it largely left off answering *why* and took up the more limited task of answering only *how*. Science was able to capitalize on mathematics as a tool and became largely descriptive instead of speculative. But this re-orientation and limitation of purpose does not adequately account for the "advantage" of scientific knowledge.

On a deeper level we can observe that science distinguishes itself from other forms of knowledge by its epistemological presuppositions. This is just a way of saying that for science the source of knowledge is different from the source of knowledge in religion or most philosophy. As we have seen, science is exclusively concerned with the empirical world of sense experience. Scientific knowledge is *about* the sensible world, *originates* in sense experience, and ultimately is *tested* against the standard of sense experience. All of this is quite obvious from our analysis of scientific method and its "levels of concern."

Other kinds of knowledge ordinarily do not share this empiricist epistemology. Religious knowledge, for example, is primarily *not* empirical in purpose, origin, or test. It often has as its purpose to describe non-physical entities (*e.g.*, spirits). It originates frequently from direct revelation to men by God (*e.g.*, in the Bible or Koran). Its standard of truth is often not sense experience, but faith or even the self-authentication of revelation. Another kind of knowledge is that based on authority. This is frequently the kind of knowledge most of us have. We have never seen, heard, or touched Julius Caesar, but we "know" he existed and was emperor of Rome. We don't even usually experience that bodies fall at the same speed despite differences of weight. In most cases what we know is based on our trust in the authority of a teacher or a textbook. Finally, another kind of knowledge may be based on intuition. For example, how can you be sure that the laws of logic hold true? The philosopher Descartes even used such a principle in judging which ideas he could be sure were true. He claimed that those which were both "clear and distinct" were evidently true. The point is that there are obviously other epistemological presuppositions besides the empirical one of science.

But does this foundation of scientific knowledge in sense experience really give science an advantage, or does it just make science different? Most people *believe* that knowledge based and tested in the "facts" of sense experience is superior. But, of course, that belief is largely a result of our living in an age when science is held in high regard. Yet why it is held in high regard is exactly our question. Perhaps the most common response is that science, with its basis in sense experience, *is* superior because it is *objective*. "All these other kinds of knowledge," it is said, "are too subjective." Only science transcends opinion. Let us look at this suggestion on a still deeper level.

What exactly is this objectivity to which science attributes its certainty and superiority? We will investigate two components which may be said to characterize objectivity. They are 1) Inter-subjective testability, and 2) Universality. Together, these give scientific knowledge its attractiveness and power.

1. One way to define 'objectivity' is to say that it is 'non-subjectivity' and then to try to define 'subjectivity.' Often, when this approach is taken, a wide gap is established between what is subjective and what is non-subjective (i.e., objective). Science is said to exclude the personal involvement, and the opinion which is said to characterize subjectivity. This kind of thinking only accentuates C. P. Snow's "gap." But is this dichotomy accurate? Is there really a kind of knowledge in which there is no personal involvement? Can the object of knowledge really be independent of the knower? Does science really have this kind of objectivity?

It is obvious that any object which is *completely* independent of the knower, i.e., related in *no* way to the knower, simply cannot be known. Furthermore, we argued in Chapter One that facts do not speak for themselves. To have knowledge, the subject must select and choose among experiences; in short he must contribute subjectivity. All data are to some extent "theory-laden," i.e., interpreted by a subject. The processes of measurement and the language of expression shape the sense experience we gather.[1] We have also observed the important role of subjectivity (imagination) in discovery. Subjectivity is present in confirmation too, to the degree that the process is dependent on observation.

It must be admitted then that pure "non-subjectivity" is not a feature of any kind of knowledge, and that there is probably a continuum, not a gap between this extreme and that of pure and arbitrary subjectivity. Rather than defining objectivity in terms which make it

94

exclusive of all subjectivity, let us focus instead on 'intersubjective testability.' We can then suppose that science will fall closer to the objective end of such a scale because its knowledge will be intersubjectively testable.

Because science is conducted in the context of a scientific community, its findings must necessarily be open to investigation and test by other members of the community. This intersubjective testability precludes the acceptance of knowledge which arises from the arbitrariness and capriciousness of the individual scientist. While, as we have seen, all knowledge must be subjective or personal, to a certain extent, the subject's involvement in this community allows the knowledge he produces to transcend his private idiosyncrasies. As I. Barbour so aptly puts it;

> Science is thus *personal but not private....* It is *personal involvement in community,* not lack of involvement, which here preserves a valid aspect of objectivity.[2]

2.   Another way to define objectivity is to distinguish between repeatable and unique events. With this distinction, objectivity might be characterized as belonging only to those types of knowledge that are limited to repeatable events. Unfortunately, this approach, like the first, also polarizes Snow's "two cultures." It suggests that while religion or history, for example, are concerned with unique events, science is objective because it deals only with repeatable ones.

But a close look reveals that this dichotomy, like that of subjectivity/non-subjectivity, also collapses. What exactly is a unique event? If one adopts a minimal definition, that uniqueness is dissimilarity in *some* respects from all other events, then *every* event is unique and not even science is objective. But if one adopts the maximal definition that uniqueness is dissimilarity in *all* respects from all other events, then *no* event is unique and all disciplines are equally objective.[3] Thus the dichotomy of repeatable and unique events breaks down also to a continuum. As Barbour puts it,

> There are thus not two different kinds of events, [repeatable] and unique. There are simply different kinds of interest in events which can be considered *both as [repeatable] and unique.*[4]

We cannot say then that science is objective in the mistaken sense that it is concerned exclusively with repeatable events. But we can say that it is closer to the objective end of a continuum because it is concerned with events *in their universal or repeatable aspects.* Science escapes the subjectivity of dealing with the peculiarities of each event, by focusing only on those features which it shares universally with other events.

These two approaches to objectivity allow us to characterize science as objective and to see that scientific knowledge, while perhaps not superior in every way, does have important advantages in some areas which give it considerable power and an aura of certainty. Once the scientists among you have taken a justifiable bow, let us turn to the other side of the coin. In what ways are scientific ideas limited?

SCIENCE'S UNCERTAINTY

Notwithstanding everything that has been said in the previous section, it is important that science not be worshipped. When science becomes scientism, and seeks to foist its methods onto all areas of human endeavor, it has over-extended itself. In considering the limitations of science, we shall consider them in two groups. In the first section below we will examine those uncertainties which arise because of the way scientific ideas are *formed*. These we shall refer to as 'Methodological Uncertainties.' In the following section we will consider those uncertainties which arise from the way scientific ideas are *structured* and *used*. These we will call 'Intrinsic Uncertainties.' While this distinction is useful, it is by no means mutually exclusive. Some of the examples used might very well fit into both categories.

Methodological Uncertainties

1.   Error — The first source of uncertainty that arises in the process of doing science is that of error. Human sense organs are prone to making mistakes and these mistakes introduce uncertainty in the theories of science. A meter is incorrectly read, read upside down, or read while switched off. A romance or stomach ache makes the technician enter a '6' instead of a '9.' A glandular secretion makes a dark plum taste sour. Any such mistake introduces error.

That people make mistakes is obvious, but *why* they make mistakes is not. Fatigue, preoccupation, physiological defects—all these contribute to making errors in observation. Perhaps error is an unavoidable result of man's capacity to do many *different* things. Error would be inversely proportional to man's flexibility. On this view, a machine designed for only one job would make fewer mistakes at that job than a machine designed to do many different tasks would make at that same job. For whatever reason, direct observation, is not completely reliable.

This unreliability affects science because science touches the level of observation in forming and confirming theories. At either point, mistakes of observation affect the theories involved. Because we are aware of this tendency we must always allow for this possibility to have occurred in any given theory. This allowance is an admission of those theories'uncertainty.

Fortunately, it has been determined that large mistakes are much less frequent than small ones and that the distribution of likely errors can be mathematically predicted. In short, there is a theory of observational error. Like most theories it does not answer *why* the mistakes are made, it only describes how they are made. The Gaussian distribution is perhaps best known for its application to results of testing where it is also sometimes referred to as the 'Bell curve' or just 'grading on the curve.' Count the dots in Fig. 1, and find your place on the curve.

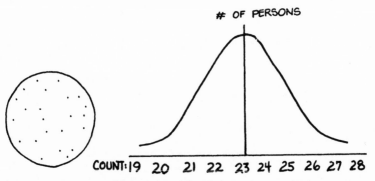

Figure 1: Error Distribution in Counting Dots

Although the mistakes made in any given case cannot be predicted, the curve gives a likely guess at how errors will usually occur. In a sense then, while errors will always occur, this theory allows us to define quite precisely the degree of uncertainty which such errors are likely to introduce.

2. Confirmation — A second, and very important, source of uncertainty in science is the process of confirmation. While the two uncertainties treated in this section are subtle, they are perhaps the most significant of all those we will discuss. A summary of the first uncertainty is that the logic of confirmation (induction) introduces uncertainty in the theories it confirms because of a vicious circularity we will call the'Problem of Induction.' The second uncertainty in confirmation is captured by what is called the 'Ravens Paradox.'

a. From Chapter One we know that the certainty of a theory is a result of what happens in the process of confirmation. We observed that prediction guides us in performing relevant experiments and that the results of these experiments are basically of two sorts.

In the first place, the experiment may show a prediction to have been false. In this case we reject the theory from which the prediction was deduced. It is much easier to falsify a theory than to confirm it. The logic of this rejection or falsification is straightforward. If the theory under test is A and the experimental result predicted is B, from our knowledge that prediction is a deductive process, we may say that, A→B. Now if the experimental result was actually not B (-B), we may justifiably argue that theory A is not true (-A). This may be summarized and illustrated as follows:

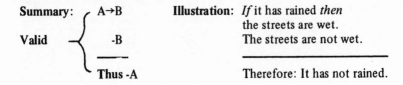

Summary: A→B   **Illustration:** *If* it has rained *then*
                                              the streets are wet.
Valid   -B   The streets are not wet.
                 ———
                 Thus -A   Therefore: It has not rained.

Figure 2: Logic of Falsification

The logic of falsification is valid *deduction* as pointed out in a note to Chapter One.[5]

The second possible result of the experiment is that it may agree with the prediction made. In this case we say that the theory has been confirmed. We were careful, however, to avoid saying that the theory had been proven and to point out that this confirmation process, while lending greater and greater certainty, must always be cyclic and never-ending.

To see why confirmation does not *prove* a theory, we need only to look at the logic of this process. If, once again, the theory under test is A and the experimental result predicted is B, we say, from our knowledge that prediction is deductive, that A→B. Now if in this case, the experimental result *is* B, (as predicted), we are *not* justified in arguing that theory A *must* be true. This would be fallacious thinking.[6] This mistake can be summarized and illustrated as follows:

Summary: A→B   **Illustration:** *If* it has rained *then*
                                              the streets are wet.
Invalid   B   The streets are wet.
                 ———
                 Thus A   Therefore: It has rained.

Figure 3: Logic of Confirmation – Invalid Deduction

Notice from the illustration that the streets could be wet from the street cleaner having passed, so wet streets do *not* prove the theory that "it has rained" although wet streets *are* a valid prediction from that theory. From this we can see that the logic of confirmation is not valid deduction, and thus cannot *prove* the theory in question.

Now we understand *why* the logic of confirmation is *inductive* as we learned in Chapter One. To further understand why induction makes confirmation a cyclic process and more importantly how it introduces uncertainty in the theories it confirms, we must look more carefully at the process of induction itself.

The process of induction, we recall, is most commonly the process of using particulars to support a general statement. Every dark sweet plum adds support to the theory that "All dark plums are sweet." Every boring lecture you hear adds support to the theory that "All lectures are boring." But the process is always cyclic and unending, never complete and certain, because there will always be *unexamined* cases, especially in the future, about which we cannot be sure. It only takes one or at most a few, such cases to falsify our theory, so the unexamined cases are significant.[7] From this alone we can see how the problem of unexamined cases makes confirmation never-ending and consequently introduces uncertainty about all theories in science.

But induction is used nevertheless in confirmation. Repeatedly successful experiments are still said to render the theory in question progressively more certain. How does science justify this use in the face of the problem of unexamined cases? One can describe the problem in terms of the diagram below by asking, "What really holds up the bridge of induction that leads from successful experiments to confirmed theorries?"

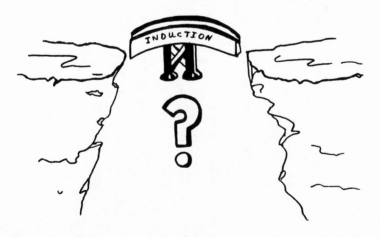

Figure 4: **What Supports the Bridge of Induction?**

To overcome the problem of unexamined cases and support the bridge of induction a simple assumption is necessary. It is that: Unexamined cases will be like the examined ones; or especially that Tomorrow will be like today. This is the assumption of the Uniformity of Nature. For example, with a few dark plums found to be sweet, and this assumption, I *can* be sure that my theory is correct. With a few sunrises in my experience *and* the assumption that nature is uniform, I *can* be sure of the theory that: The sun rises every morning. The Uniformity of Nature supports the inductive bridge and justifies its use in the confirmation process.

But an obvious question arises: What supports the Uniformity of Nature assumption? Because it is just an assumption, it too must be justified. This is a very important question. After all, the very certainty of scientific theories comes from confirmation which is an inductive process. But induction rests on this assumption so we must be sure that it is a justifiable one.

Figure 5: **The Classic Problem of Induction**

If we analyze the kind of thinking that leads us to assume that nature is uniform we discover that it is inductive in nature. Suppose I were to ask you why you believe that nature is uniform, or why you believe that the laws of nature tomorrow will be pretty much the way they are today. You might very well answer by saying, "Well nature always *has* been uniform," or, "I've never known it to be otherwise, so I don't expect it to change tomorrow." But this is just another case of using particulars (and especially the past particulars of your own experience) to support a general statement (one that especially applies to the future). This is an induction. But then what have we done? We have tried to argue that induction is supported by the Uniformity of Nature assumption which in turn is justifiable only by another induction. That kind of reasoning is a vicious circularity and is not acceptable. The inductive bridge of confirmation collapses. This we call the 'Classic Problem of Induction' and it can be illustrated as in Figure 5. In the diagram the steel chains illustrate the circularity involved. They certainly would not keep the bridge from falling.

The Problem of Induction is perhaps best summarized and its point made obvious by the story of Theobald the Turkey.

"Theobald the Turkey
and the Uniformity of Farmer Greenjeans"

Theobald the turkey was born one spring day on a farm in Ohio. Now every morning, Farmer Greenjeans came with a bucket of golden grain. Theo found that by rushing ahead of the others, he got the best kernels and more to eat. After several months of this, however, he was concerned to see that the older turkeys always hung back and that the oldest stayed in the *very* back. They warned constantly that Greenjeans was unpredictable and inconsistent. But Theo only scoffed: "That's just religious superstition you old buzzards. Look how fat I am. Every time I run ahead I only get more food." He even tried convincing other turkeys of his theory. "I've tested it scientifically," he said. "I've tried it 99 times. When I run ahead I get more food. It's only reasonable that it will work again."

When the next morning dawned, Farmer Greenjeans was right on time. Fat Theo dashed eagerly far ahead of the others. What Greenjeans carried was not, however, in a bucket, but glistened over his shoulder. It was the fourth Thursday of November.

In the final analysis there seem to be only two options open. Either we take the Uniformity of Nature assumption on faith or reject the entire scientific method, and especially the all important justification of theories by confirmation.[8] This consequence applies equally to our daily concerns as well, insofar as science and common sense are similar. There is no reason apart from faith for you to believe the sun will rise tomorrow; it is not even rationally probable.[9] Science cannot give you that reason. The certainty of its laws or theories rest on faith. Practicing scientists either accept this assumption consciously on faith or else ignore the issue and accept the assumption unconsciously as witnessed by their continuing to work.

From this discussion we conclude that there is a subtle but very profound uncertainty in the theories of science which results from the method by which they are formed. The Problem of Induction seems inescapable, and science, not unlike a religion, appears ultimately to rest on faith.

b.   A second difficulty which arises in connection with confirmation is the   Ravens Paradox. [10] Suppose the theory (T) we wish to confirm is that:

T: All ravens are black.

In symbols we can represent this as R→B! Now according to our understanding of confirmation, we would look around for ravens to see if they are black. Each black raven would be a positive instance of our theory and would thus further confirm it.

But according to the rules of logic, statements, including our theory T, have certain fixed properties. In particular, whenever a statement is true, its   *contrapositive*   must also be true. In fact the two are taken to be identical for all logical purposes. For example, the contrapositive of our theory would be:

T': All non-black things are not ravens.

Because these two statements (T and T') are held to be logically identical, it follows that anything which confirms one will also confirm the other. But anything which is not black *and* not a raven will be a positive instance of T' and thus will confirm it; for example, the Washington Monument. The paradox then is that things like the Washington Monument must also be taken as confirming our theory T that all ravens are black. The problem is only made worse when other logical equivalents to T are examined. We discover that *anything* which is not a raven and *anything* which is black will confirm our theory. The paradox is captured in Goodman's phrase: "The prospect of beginning to investigate ornithological theories without going out in the rain is so attractive that we know there must be a catch in it."

While this particular problem has received much attention in the philosophy of science, and can probably be avoided, it is typical of the kinds of problems facing investigators into the logic of confirmation. When taken together with the Problem of Induction, such questions reveal that considerable difficulty arises in trying to justify the confirmation process. This difficulty must surely be reflected in the theories of science which all depend upon this process for their certainty.

3.   Presuppositional — There is a final topic which deserves consideration in this section on methodological uncertainty. Because it is equally applicable to *all* forms of rational inquiry and not just to the scientific method, it is better not to refer to it as a source of uncertainty but rather only as a limitation. In brief it is the limitation imposed on ideas by the presuppositional character of thought. Whether it

results from the physiological structure of the brain, the structure of language, or something else, the products of human thought are always limited by the fact that they reflect presuppositions that perhaps could have been otherwise.

A simple way of stating this fact is that in science as in any field, the answers one obtains are shaped by the questions one asks. We have already seen the truth of this fact in our discussion of explanation types. Suppose for example that a young man commits a serious crime. Suppose then that his case is studied by a variety of persons who consequently make reports to a judge. The judge might easily be baffled by their reports believing that surely some mistake had been made and that each was describing a different young man and a different event.

A lawyer would report on the young man and his activities in the clinical terms of what acceptable evidence there was on which to prosecute, what precedent there was for this case, and so on. He would pick out and focus on items such as whether the young man had been informed of his rights at the time of arrest, or whether he had been permitted a phone call. A social worker, on the other hand, might report on the young man's family life, economic condition, location of residence, and job status as the true causes of his crime. A psychiatrist might describe his childhood or his emotional state at the time of arrest as evidence of an unbalanced and thus non-responsible personality. A clergyman would perhaps see the young man as individually responsible for an action that was not only illegal but immoral. He might believe that justice was in order but might also plead for lenience in punishment. And finally, the young man's mother might refuse to see the crime at all because she saw only a boy who was basically good, though occasionally made a little mischief—like all boys his age. The obvious point is that depending on our presuppositions—i.e., the questions we ask—the answers we get will be different. This is true of science and the questions it asks.

Another way of saying the same thing is that we must always begin *somewhere* in approaching a problem. In Chapter One we saw that the "ants" got nowhere in discovery because they didn't know where to begin or which facts were important. The facts didn't speak for themselves. It was necessary to launch out and create a framework by which to sort out which facts were relevant. This is not unlike what the various reporters mentioned sought to do. Each had his own set of presuppositions, his own framework by which to sort out the facts of the case. In science, it is hoped that the working hypothesis which serves as the starting place will ultimately be justified by the standard of sense experience. Yet this very empiricist bias is itself a presupposition of science. Can *it* possibly be justified by something else? Even if so, there must ultimately be a starting place which cannot be further justified.

This is normal, yet it reminds us of the limits of our knowledge—even of knowledge in general—for our presuppositions could perhaps just as easily be different. Our only recourse is to be critical of our presuppositions and open to alternatives.

When this self-criticism is absent or superficial, it often results in theories which while perhaps not wrong, are severely limited by the *obviousness* of their alternatives and the apparent arbitrariness of their presuppositions. In the heated debate on evolution, for example, the record of fossils is important. The dating of these fossils has been used as evidence that earlier species were less sophisticated than later ones into which they are said to have evolved. Of course the dating of these fossils is crucial. One method for this dating uses a group of fairly common fossils as reference or "index" fossils, by which all others are dated. These index fossils are themselves dated, it seems, largely by ranking them in order of complexity and assigning relative ages respectively. But it is apparent here that the ascending order of complexity has been *presupposed* in interpreting the data and hence cannot be *concluded* from that data as well. This is circular thinking. While perhaps overly simplistic, the example shows that many other presuppositions could have been used in sorting the data, some of which might have led to opposite conclusions. This is not to say that the one used *is* mistaken, only that the value of the evidence is severely limited by the obviousness of other alternatives and the apparent arbitrariness of its choice.

Intrinsic Uncertainties

In the following two sections we will consider sources of uncertainty which stem from the structure and use of the scientific ideas in themselves. It must be remembered, however, that the distinction of Methodological and Intrinsic Uncertainties, while useful, is not mutually exclusive. Particularly in the second section below, there is a strong sense in which that uncertainty may be said to arise from the *methods* of science.

1.　Statistical — It was suggested in Chapter Two that the theories of science are only rarely of the ideal universal form, "All A's are B's." In reality the theories and laws are of a statistical nature like the statement that "80% of all cigarette smokers get lung cancer." This statistical character of such laws leads to at least two sources of uncertainty in their use.

a.　In the first place, such laws are frequently just statistical correlations; statements describing things which have been observed frequently to go together or correlate. These statistical *correlations* are often confused with causal *connections*. But just because two things *go* together doesn't mean that one causes the other.[12]

For example, a "scientific" study might reveal that "90% of all executives drive cars costing more than $10,000." You might conclude that if you want a luxury car you should become an executive, or that executives buy luxury cars because they are executives, or even that if you want to become an executive you should buy a luxury car. (While this last may be laughable, it appears to be at the heart of *much* modern advertising—and it works!) Or as another example, another study might reveal that "90% of those earning over $50,000 a year also smoke." From this correlation you might conclude that their smoking causes (at least in part) their riches, or even that their riches cause them to smoke. Finally, research may reveal that "90% of all smokers get lung cancer," and you may infer that smoking causes cancer.

The problem is to be able to know which statistical correlations do actually reveal direct causal connections (and then which is cause and which is effect), which reveal indirect causal relations, which

represent mutual effects of another cause yet unknown, and finally which are merely accidental correlations with little if any connection. The *point* is that this difficulty of knowing how to interpret statistical correlations is what introduces uncertainty in their use. Because scientists do so frequently use statistics to form their generalizations, the danger must be recognized. A final illustration emphasizes the hazard.

> A group of non-drinkers from the "Tea! Totally" organization raised funds at a "Tea-Off" to conduct a scientific investigation of what makes those abominable mixed drinks so intoxicating. They carefully collected samples and began their study. They observed that Gin and soda water was intoxicating. They also noticed that Rum and soda water was intoxicating. Finally they observed that Whiskey and soda water was intoxicating. After careful statistical correlation of the results they soberly announced that soda water caused intoxication.

      b.   A second uncertainty which arises from the statistical character of many scientific laws is that their application to particulars is frequently impossible. Statistical laws by their nature apply to aggregates or classes of events or people. For example, in physics there are half-life decay laws which describe the rate of natural decay for radioactive elements. They are stated in terms of the time required for *half* of the radioactive atoms present at any moment to decay. If the half-life is 10 minutes and there are 1 million radioactive atoms present now, then in 10 minutes there will be only 500,000 and 10 minutes later only 250,000 and so on. While these laws are useful when applied to large numbers of events, they are completely uncertain in predicting when any *individual* radioactive atom will decay.

      Perhaps more obvious still is the limitation of Census Bureau statistical "laws." Suppose that in 1971 the Bureau announced that the average American family had 2.3 children. Could you find the "average family" referred to? And if you could, would they actually have 2.3 children? The point is that with statistical laws, uncertainty regarding particulars is unavoidable.

      2.   Heisenberg — A final source of uncertainty arises from the heart of modern physics. The 19th century physics of Newton was among other things ultimately deterministic. By this is meant that the entire universe was rather like an enormous machine whose behavior could in principle be completely predicted. Newton thought of the world as consisting of irreducible bits of matter (atoms) which were in motion. These motions always conformed completely to Newton's own three laws of motion.[13] The physicist Laplace carried this world-machine

view to the extreme. He argued that if a superintelligence could know the positions and velocities of all particles in the universe for only one instant, then with the help of the laws of motion it could see the past and future as though it were present.

> We sought then to regard the present state of the universe as the effect of its anterior state and as the cause of the one which is to follow. Given for one instant an intelligence which could comprehend all the forces by which nature is animated and the respective situations of the beings who compose it—an intelligence sufficiently vast to submit these data to analysis—it would embrace in the same foundation the movements of the greatest bodies of the universe and those of the lightest atom; for it, nothing would be uncertain and the future, as the past, would be present to its eyes.[14]

Laplace's "intelligence" sounds frightenly like a very large computer. In fact his "intelligence" is more commonly referred to as Laplace's "Demon."

But with the dawn of contemporary physics—only 80 years ago or so—this determinism quickly disappeared. Instead of exact values for prediction, the new Quantum Mechanics could only give probabilities with an uncertainty that varied greatly. It was shown that certain pairs of concepts (for example velocity and position) describing any given particle, were related to one another in a peculiar way. The more accurately you knew one of them, the less accurately you could predict the other, and vice versa. For example, the more accurately you measured the position of an electron in an experiment, the greater the uncertainty regarding its velocity. This is Heisenberg's Principle.[15] Taken together with the wave-particle dualism and its Principle of Complementarity, Heisenberg's Principle has upset Newtonian physics and revolutionized its world-view.

Unfortunately, Heisenberg's Principle has been much abused. It has been frequently invoked to justify everything from human freedom or idealism to an interventionist theory of miracles. Such abuses are born of a failure to investigate carefully the physics involved and its possibile interpretations. For our purposes, it is obvious that nearly any interpretation introduces a significant *intrinsic* uncertainty into the theories of physics and thus indirectly into the theories of all other sciences of nature. Nevertheless, let us consider some possible interpretations.

Some argue that the uncertainty is only a result of temporary human ignorance. Among those who adopted this interpretation were Einstein and Bohm, and more recently DeBroglie. These argue that there *are* detailed subatomic laws which are rigidly deterministic. They believe that we do not presently know those laws but that some day they will be discovered and exact prediction will once again be possible—at least in principle—and the resultant uncertainty of Heisenberg's principle will simply disappear. Einstein summarized the convictions of this interpretation, uneasy with a world of intrinsic uncertainty and chance, in the famous phrase: "My God does not play dice."

A second interpretation held by Bohr is that Heisenberg's uncertainty arises from unavoidable experimental or conceptual limitations, which forever keep us from knowing whether the things we observe are determinate and predictable or not.

One version of this second interpretation attributes the uncertainty to an intrinsic experimental difficulty: the uncertainty is introduced by the process of observation. Suppose that the CIA wished to obtain precise (certain) information about the economic conditions of the country of Bogo Bogo, deep in the heart of Africa. To this end it

obtained a spy who was fluent in Bogoese, trained him in underground work, and parachuted him by night into Bogo Borough, the capital city. They then manned the shortwave for their first reports from the agent. Suppose, however, that when our agent awakened the next morning he discovered that the entire population of Bogo Bogo lived in Bogo Borough and consisted of only three persons. His reports would obviously be of little use to the CIA, because of the effect his presence would have on the very thing he was sent to observe. Finding that their population had jumped 33% overnight the mayor of Bogo Borough opens a new Bogo Chef restaurant that same day. By the next day the two deputy mayors had both opened competing Bogo King franchises. Our agent sent his report that the economy was booming and then returned to the U. S. Of course the booming report was probably inaccurate, and after his departure it is hard to say what became of that economy. In short, his presence as an observer disrupted the economy, and introduced uncertainty in the reports he gave his superiors. The same difficulty would be present even if Bogo Bogo had a population of 1 million, though obviously the uncertainty in the reports would be so small as to be *negligible for most purposes*. The uncertainty would remain nonetheless.

This intrinsic experimental limitation affects every attempt in science to measure anything. To observe the position of an electron for example, one must "send in an agent," a light beam. The light disturbs the electron's position in measuring it and thus introduces uncertainty in the result obtained. In noting the position of a box in the basement, you disturb its position by shining a light on it. There is uncertainty in your measurement, although in this case, the light disturbs such a huge object so little as to be neglible. Note that the disturbance is introduced by the observing process, and not by the observer's mind, and that it is unavoidable. The resultant uncertainty is an intrinsic limitation of human knowledge which permanently prevents us from knowing whether the events we observe are determinate or not.

A second version of this interpretation attributes Heisenberg's uncertainty to an intrinsic and thus inescapable conceptual limitation. Man himself established the categories (concepts) to be imposed on the events we observe. *We* decide that particles have position and velocity instead of "posicity" or "velotion," and perhaps the categories we've chosen just don't fit. It wouldn't work well to describe oranges in terms of charm or triangularity. On this view, the uncertainty of Heisenberg's principle arises because our choice of concepts must inevitably always fail to really fit the events we observe. This misfit introduces uncertainty into the measurements we do obtain.

Both the first and second versions of this interpretation of Heisenberg's uncertainty leave us agnostic about whether nature *itself* is indeterminate. We can never know because intrinsic experimental and/or conceptual limitations forever introduce uncertainty into scientific knowledge.

A third and final interpretation of Heisenberg's uncertainty is held by Heisenberg himself. It is the straightforward view that the uncertainty which appears in observations and thus in science in general arises from the fact that nature *is* in itself uncertain. This is obviously the extreme opposite of the Laplacian determinism whereby the world is likened to a completely predictable machine. It is the view that at any given moment there is not one but many different possible futures. The uncertainty of science is not a result of a temporary ignorance nor is it the result of measuring or conceptualizing, but arises from the uncertainty, indeterminism, and lawlessness of nature itself.[16]

Each of these three interpretations has a very different view of nature. The first takes the world to be deterministic and man's knowledge to be only temporarily uncertain. The second takes man's knowledge to be intrinsically and unavoidably uncertain while the true nature of the world will be forever unknown. The third view states that man's knowledge is unavoidably uncertain *because* nature is in fact indeterminate and uncertain.[17] But, in summary, each of these three interpretations agrees that there is a fundamental uncertainty in the theories of science about nature.

SUMMARY

In this chapter we have been concerned with the limitations of scientific ideas. Our guiding question has been: How certain can we be of the theories which science *produces* and *uses?* The main point was "to provide a healthy balance of appreciation and scepticism regarding the claims of science" in order to help overcome the gap between Snow's "two cultures."

In the second section we examined some of the bases on which science can claim its advantage of certainty: specialized questions, empiricism, and most importantly, objectivity. The strength of science seems to be especially in its "objectivity" when that concept is carefully defined and limited.

In the third section we undertook what is perhaps the more urgent task of pointing out how scientific knowledge contains *much* uncertainty which is all too often overlooked by both "priests" and "believers" of this popular "religion." Those uncertainties fell loosely into two groups; those which arise from the way scientific ideas are *formed* and those which arise from the way scientific ideas are *structured* and *used.*

The moral is that while science and scientists must not be

scorned, their work must always be kept in the proper perspective of its many limitations and uncertainties. The scientist must be proud, but not haughty, and the layman must be respectful, but not worshipful.

APPENDIX

CHAPTER TWO

The Development of Scientific Ideas:
The History of Science

I. Introduction
    - A. Review: The Snow Gap
    - B. A Distinction: Formation vs. Development
    - C. A Misconception: Development is Cummulative
    - D. Preview: Development is Evolutionary and Revolutionary

II. The Development of Scientific Ideas of Motion
    - A. Aristotle
    - B. Impetus Theorists
    - C. Galileo
    - D. Newton
    - E. Conclusion
        1. Gestalts occur
        2. Anomalies encourage Gestalt thinking

III. The Development of Scientific Ideas In Astronomy
    - A. Aristotle
    - B. Ptolemy
    - C. Copernicus
    - D. Galileo
    - E. Conclusion
        1. Three theory selection criteria
        2. Significance of personal and social factors

IV. The Development of Scientific Ideas: Evolutionary or Revolutionary?
    - A. Question
    - B. Evidence
        1. Evolutionary: General/Anticipations/Modifications
        2. Revolutionary: Motion/Astronomy
    - C. Conclusion: Two Versions of a Balance
        1. Kuhn
        2. Toulmin

CHAPTER THREE

The Nature of Scientific Ideas:
Theory Structure

I.   Introduction: A Basic Question

II.   General Characteristics of Theories

III.   What *Are* Scientific Ideas? Some Distinctions

     A.   Theories, Laws, and Formulae

          Two Approaches:

          1.   By degrees of Confirmation

          2.   By generality and components

          An Illustration

     B.   Laws of Nature versus Accidental Generalizations

IV.   What do Scientific Ideas *Mean?* Two Interpretations

     A.   The Realist View

     B.   The Instrumentalist

V.   Summary

CHAPTER FOUR

The Use of Scientific Ideas:
Explanation

CHAPTER FIVE

The Limitations of Scientific Ideas

I. Introduction
    A.   Purpose and Main Point
    B.   A Capsule History of Science's "Reputation"

II. Science's Certainty
    A.   Specialized Focus
    B.   Empirical Epistemology
    C.   Objectivity
        1.   'Non-Subjectivity'
        2.   'Repeatability'

III. Science's Uncertainty
    A.   Methodological Uncertainties
        1.   Error
        2.   Confirmation:
            a.   Problem of Induction
            b.   Confirmation Paradox
        3.   Presuppositional
    B.   Intrinsic Uncertainties
        1.   Statistical Uncertainty
        2.   Heisenberg Uncertainty

IV. Summary

NOTES

INTRODUCTION

1 C. P. Snow, *passim.*
2 It seems that as science is blamed for many of the serious environmental and social problems we face today, there has been a growing disillusionment with the infallibility of science. This was especially true in the late 1960's and early 1970's.

CHAPTER ONE

1 Religion, philosophy and many other areas of human concern can also be thought of as both dynamic and static.
2 T. H. Huxley, pp. 42-129 *passim.*
3 W. J. Sinclair, *passim.* Cf. C. G. Hempel, (1966), pp. 3-6.
4 I. Newton, (1972). Cited in J. Bronowski, p. 226. Cf. I. Newton, (1952).
5 *Ibid.*
6 *Ibid.*
7 The use of these generalizations is the purpose of science as a static collection of theories. See Chapter Four for discussion.
8 J. T. Hardy, p.
9 T. H. Huxley, pp. 42-129 *passim.*
10 The latter two are often combined and referred to as the 'Justification' or 'Verification' phase of scientific method. They are distinguished here to reveal the distinct thought processes involved. The discovery phase is alternately referred to as the 'Invention' process.
11 This use of the word 'discovery' conforms to its use in the literature of the philosophy of science, but probably differs from the more common use which is limited to discoveries which *are true.* The sense in which a theory "accounts" for a problem—i.e. explains it—is dealt with in Chapter Four below.
12 As will become evident below, the "logic of abduction" can be characterized generally, but not analyzed into progressive mechanical steps. In this sense it *lacks* the logical character of either deduction or induction. For further discussion of this issue see N. R. Hanson, (1970).
13 This term is used by C. S. Peirce to distinguish discovery from the inductive process of confirmation. C. S. Peirce, pp. 235-255. The term 'retrodiction' is

also sometimes used in reference to the discovery process.

[14] R. Blackwell, pp. 104-110.

[15] *Ibid.* Sometimes also called the 'narrow inductivist' or 'encyclopedic' approach. Some have said that Bacon's method is vindicated by the fact that he discovered heat to be essentially motion long before the Kinetic-Molecular theory of heat explained why this is so. Some might say Bacon was just lucky. Even if this is false, Bacon's success was probably less because of what he *said* about method than what he *did*, using imagination not a mechanical process.

[16] Popper's Fable is here paraphrased from J. Bronowski, *Science and Human Values*, (New York: Harper, 1965), p. 14.

[17] R. Blackwell, pp. 104-110.

[18] The kind of relevance/relationship which positivist philosophers of science have usually required, is spelled out in Chapter Four, where the explanatory *use* of theories is discussed.

[19] *Ibid.*

[20] R. Blackwell, paraphrasing F. Bacon, *Novum Organon*, Bk. I, par. 95.

[21] There are naturally many attempts to understand the discovery process. Among such attempts are those of R. Blackwell and A. Koestler. It is also often the concern of psychologists.

[22] There are many hidden premises left out here, *e.g.*, "You cannot eat in class," "You must eat if you are hungry," etc.

[23] This deductive necessity arises when an argument conforms to a valid argument form. The form which applies in our example is sometimes called *modus ponens* or affirming the antecedent: If A then B. But A is true. Therefore B is also necessarily true. $(A \rightarrow B, A, \therefore B)$.

[24] Of course this would not be true on a weaker sense of 'consequence' than we are using here.

[25] But as Karl Popper insists, this falsifying event must be repeatable. Furthermore, as Kuhn points out in *The Structure of Scientific Revolutions*, scientists become quite dogmatic in defending a prevailing theory during crisis periods. So falsification is not as simple in practice as this account may suggest. See also Chapter Two below, p. 51.

[26] The logic of confirmation is inductive, the logic of falsification is actually deductive and of the form called *modus tollens*. While this logic of falsification is deductively valid, the logic of confirmation (induction) is deductively fallacious. It consists of the fallacy of affirming the consequent; i.e. A implies B, B, therefore A, and accounts directly for the cyclic and unending character of confirmation. See Chapter Five below, pp. 97-102, and especially p. 98.

[27] *Ibid.* Note also that to be precise, the same problem which keeps inductions from *proving* theories, also keeps it even from making them "progressively more certain" (probable). (See B. Russell "On Induction").

[28] *Ibid.*

[29] Arthur Conan Doyle, "The Adventure of the Blue Carbuncle, Vol. I, pp. 244.

[30] *Ibid.*, p. 245.

[31]Doyle, "A Study in Scarlet," Vol. I, p. 27.

[32]This raises an interesting question about who makes the better scientist. Would it be someone who has much experience with which to quickly rule out many mistaken hypotheses but whose experience prevents them from thinking of new analogies which might be helpful? Or might it be better to have less experience and consequently fewer preconceptions to hinder free association and original synthesis?

[33]Doyle, "A Study in Scarlet," Vol. I, pp. 83-84.

[34]*Ibid.*, p. 31.

[35]Doyle, "The Man With the Twisted Lip," Vol. I, p. 240.

CHAPTER TWO

[1] See W. A. Levi, *Philosophy as Social Expression* and J. Collins, *Interpreting Modern Philosophy* both of which focus on this historiographical issue.

[2] T. S. Kuhn, (1962), p. 1.

[3] But of course, it may also be argued that the selection of these examples instead of some others, prejudices the conclusions drawn, and opens the question of whether in this chapter I commit the same distortions of history which Kuhn and I criticize in science textbooks. Part of the answer may be that presuppositions can never be eliminated in writing history. The conclusion regarding balance drawn from these two prominent episodes in history must speak for itself. Beyond this the reader must investigate the history of science for himself or herself. Cf. T. Kuhn, (1962), and H. Butterfield.

[4] Falling objects may appear to have no mover, but remember they are always moved by their attraction to the earth.

[5] H. Butterfield, p. 18.

[6] *Ibid.*, pp. 24-25.

[7] This depends on one's interpretation of 'Impetus.' If it is taken to be a fluid-like mover which permeates the object and gradually "oozes" out, the theory is Aristotelian. If impetus is simply a name which refers to a behavioral characteristic of all masses (like inertia), the theory is modern.

[8] H. Butterfield, p. 23.

[9] *Ibid.*, p. 93; cf. W. C. Dampier, p. 130, note 4; Galileo was supposedly attempting to verify that gravitational attraction *varies* with mass, pulling harder on the more massive object than the less massive. The net result would be that the two objects would fall with *equal* acceleration and strike the ground at the *same* time. Supposedly Galileo observed that they did.

[10]Galileo fell short of the full blown Cartesian or Newtonian statement of inertia because of his Copernicus-like insistence on the perfection of circular versus linear motion. But the key switch had been made by Galileo. Only the direction of natural motion remained to be resolved.

[11]Here the problem of classifying Galileo becomes as difficult as the problem of classifying the Impetus Theorists. Is there really any difference, except in words, between Aristotle's view that motion is explained by desires and the Im-

petus Theorists' view that motion is explained by a mysterious something called 'Impetus,' and Galileo's theory that motion is explained by the mysterious nature of the world, sometimes later called the 'Law of Inertia'? It is interesting to speculate whether inertia is anything different from Aristotelian entelechies (desires). If not, then perhaps much more can be said for an evolutionary view of science from this chapter in history than for a revolutionary view.

12 See R. Taton, p. 239, regarding this controversial story.

13 C. Morphet, pp. 12, 27. This is sometimes called 'Copernicus' conservatism!'

14 *Ibid.*

15 While it may seem ludicrous to us to have the stars all spin so fast about the earth as Ptolemy and Aristotle argued, this presented a far smaller problem to them than to us. Because the stars and sun were thought to be light-weight and small, for them to spin fast was far more likely than for the massive, apparently unmovable earth to *spin* on its axis. Such an idea countered every intuition of experience, reminding us again that early science was *not* foolish, given the observations available.

16 C. Morphet, p. 53. The astronomical evidence did refute Ptolemy but *not* later Ptolemaic versions of Tycho.

17 *Ibid.*, p. 26.

18 *Ibid.*, pp. 27-36.

19 *Ibid.*, p. 54. For a comparison of this "instrumentalist" view of theories with the realist view, see Chapter Three, pp. 65-69 below.

20 *Ibid.*, p. 58

21 *Ibid.*, p. 52. Astronomical data regarding craters on the moon and regarding the moons of Jupiter was unacceptable because it was obtained using the mysterious telescope.

22 *Ibid.*, p. 53.

23 No suggestion of such a reply is made in the very thorough book by G. de Santillana, *The Crime of Galileo* (Chicago: University of Chicago Press, 1955). See esp. pp. 237-321.

24 T. S. Kuhn, (1962), p. 1.

25 Pursuing the dispute, the evolutionists say no, Newton understood mass to be "time dependent" when he defined force not as ma=mdv/dt, but as dP/dt, thereby putting mass into the time function.

26 T. S. Kuhn, (1962), *passim.*

27 See T. S. Kuhn, "The Function of Dogma in Scientific Research."

28 Two additional assumptions of Darwin's theory are: 3) Malthus' assumption that there is a *struggle for existence* within any species due to the geometric growth of population and only arithmetic growth of food supplies. 4) The presence of a mechanism of *genetic inheritance* whereby favored (selected) variations can be passed down to subsequent generations; i.e., the variations which are selected are more likely to occur in subsequent generations. Tall giraffes are more likely to have tall offspring than to have short offspring. We know today that such a mechanism exists.

29 The phrase "for that time" shows the sense in which for Toulmin the

function of theories in explanation (see Chapter Four below) is "historically relative." As environments vary, the criteria for "fittest" explanation will change. See Toulmin (1961), regarding his "Ideals of Natural Order" and how they change.

[30] S. Toulmin, (1967), (reprinted in W. H. Truitt *et. al.*), pp. 106-117.

[31] This point is also made by P. K. Feyerabend in "How to Be a Good Empiricist," reprinted in Brody.

[32] S. Toulmin, in Truitt, pp. 113f.

[33] For a more complete comparison of Toulmin and Kuhn see Suppe, pp. 127-151; 633-680. It would appear from Suppe's analysis that the philosophy of science has come to the stage where the contexts of discovery and justification and use and development for theories can no longer be easily separated. The Kuhn/Toulmin approach to scientific development (Suppe's term: *Weltanschauungen* Approach) is no longer dominant over the positivist account of theoretical structure and use. Both seem to have given way in the 1970's to a broader concern for both logical and historical factors and the emergence of an historical realism regarding theories.

CHAPTER THREE

[1] In Chapter One, we specified two "levels" of concern as the playing field of science; one particular and empirical, the other general and abstract.

[2] E. Nagel refers to these principles taken together as the 'calculus' of a theory, p. 90f. They are also sometimes misleadingly called the 'hypotheses' of a theory by N. R. Campbell in *Physics the Elements.* The distinction of internal and bridge principles is found in many places including C. Hempel's *Philosophy of Natural Science,* p. 72ff.

[3] These are also sometimes called 'Rules of Correspondence' (see Hempel, [1966], pp. 72ff.), or the 'Dictionary' of a theory (see Campbell). Also see Suppe, p. 17ff. (especially note 33) for other synonymous terms.

[4] Implicit definition is provided by the internal principles which stipulate how the theoretical terms must relate to one another.

[5] C. Hempel, (1966), p. 73f.

[6] Absolute value means to take the result between the lines as positive even if it turns out negative. *E.g.* $\mid 6-5 \mid = 1$ and $\mid 6-7 \mid = 1$. Thanks go to Profs. R. Dacey and P. Derr for their help on this illustration.

[7] This last qualification prevents us from having to include all excitement as worrying.

[8] The product is measured, of course, in inch-hours per week!

[9] NTPS for freshmen is 6, for juniors is 1, for seniors is 1 1/2.

HL for freshmen is 1, for juniors is 6, for seniors is 4.

WL for freshmen is 2 (20 hours of worrying per week), for juniors is 1/3 (3.3 hours of worrying per week), and for seniors is 1/2 (5 hours of worrying per week).

[10] E. Nagel argues that theories must also contain a model and by this he means a formal model or interpretation of the calculus; Nagel, p. 90. M. Hesse

125

insists that models of a material sort are essential to theories. See her *Models and Analogies in Science* and her *The Structure of Scientific Inference*. We will look at this matter somewhat more in Chapter Four.

[11] This is the old Logical Positivist problem of determining meaningfulness using the Verification Principle. Many conclude that the effort has failed, but the distinction is useful for our purposes.

[12] For an excellent introductory discussion of Operationism, a third but currently less popular alternative, see Hempel, (1966), pp. 88-100.

[13] Nagel argues that theoretical terms distinguish theoretical laws (theories) from empirical laws. See Nagel, p. 80.

[14] See Hempel, (1966), p. 94, where his term is 'systematic import' or 'theoretical import.'

[15] A correspondence theory of truth is assumed, where the truth of a proposition depends not on its internal *consistency* or *coherence* with other propositions, but on its accuracy in *corresponding to* the actual empirical state of affairs observed.

[16] For a more complete discussion of this objection and the entire question of the cognitive import of theories and theoretical terms, see Nagel, pp. 117-152.

[17] See Suppe, pp. 80ff. and P. Achinstein, (1968),

[18] Nagel, p. 130, cf. note 13 above.

[19] Nagel concludes that the difference between these two views is only a matter of words; Nagel, p. 152.

CHAPTER FOUR

[1] The words 'finished' and 'static' are in quotes for at least two reasons. First, of course no theory is ever completely confirmed as pointed out in Chapter One and as developed in Chapter Five below. Second, most theories undergo development and are changed with time. Whether this development is evolutionary (and the change gradual) or revolutionary (and the change radical) is subject for Chapter Two on the "Development of Scientific Ideas." Considerable material on this subject may be found in T. S. Kuhn's book, *The Structure of Scientific Revolutions.*

[2] See Chapter One, p. 13 above.

[3] Illustration No. 2 and the idea for such illustrations is drawn from Nagel, pp. 15-20.

[4] These are sometimes referred to as the 'explicandum' and 'explicans' respectively. For either pair of Latin words, they can be conveniently distinguished by remembering that the ending 'ans' on 'explanans' (or 'explicans') identifies the *ans*wer not the question.

[5] There are many ways to classify explanation types. Nagel uses a fourfold method. I have omitted his 'genetic' explanation because of its similarity especially to the teleological types and because of the limited scope of this discussion.

[6] Notice this characterization is not intended to imply that all modern theories are to be interpreted by the Descriptive view of cognitive significance. Although many theories have been interpreted thus, equally many have been taken instrumentally. Even if taken instrumentally, or realistically, there remains a sense in which the theory can be said to *describe* experience by focusing on behavior rather than purpose. This is the point being made here. Cf. Chapter Three, p. 58, above.

[7] For more discussion of the role of analogy and models in explanation see Nagel, pp. 90-97 and 107-117. Also see Hesse, (1970), pp. 157-177.

[8] Holton and Roller, *Foundations of Modern Science*, p. 160 (reprinted in Hempel, [1966], p. 48).

[9] See especially C. Hempel and D. Oppenheim, (1948), and Hempel, (1965).

[10] See note 7 above.

[11] See note 6 above.

[12] Of course this is an overly simplified form of that law even as applied to earth. More precisely, Newton's law is $F_{grav} = gm_1 m_2/r_{12}^2$. The explanans usually contains such covering-laws and certain *initial conditions:* here the statement that the apple was unsupported.

[13] Notice too that one must be careful in defining scientific laws that one does not allow universals that are merely accidental or that are tautologies. "All the people in this room are under 40 years of age" does not count as a scientific law even if it is a universal statement and even if it is true. Its truth would be an accident of circumstance, not a general feature of nature. Nor does the statement "All bachelors are single" qualify as a scientific law, just because it is a true universal statement. It is a statement about the meaning of words, not about nature. See Chapter Three, pp. 64-65 above.

[14] Put in other words, the explanandum is *not* a necessary result of the explanans. I-S explanation is not to be confused with what might be called D-S explanation. A D-S explanation would retain the deductive relation between explanans and explanandum, but would have statistical premises and a statistical conclusion. *E.g.:*

(1)   85% of all smokers get cancer
(2)   Joe is a smoker

Therefore   Joe has an 85% chance of getting cancer

[15] For more detailed treatment of the I-S model and the problem of statistical ambiguity, see Brown, pp. 58-60, and Hempel, (1965), pp. 394ff.

[16] That is to say, we must carefully distinguish between the question of the degree of support for the *premises,* and the question of the degree of support for the *conclusion* in any covering-law explanation argument. Or again, we must distinguish the certainty of the covering-laws themselves from the certainty of the connection of such laws to the explanandum. Thus explanations are I-S when probability figures in the *latter,* not the *former* matter (where probability usually

pertains anyway because of the confirmation process. See Chapter Five).

[17] See Brown, pp. 53-58; and Hempel, (1966), p. 249.

[18] But even Hempel concedes this in *Aspects*, (1965), p. 368.

[19] As Toulmin would say, they had "foresight," but no "understanding." See Toulmin, *Foresight and Understanding*, (1961).

[20] Used by P. Achinstein, (1971), p. 105, citing Arthur Collins.

[21] See *e.g.* Hempel, (1965), pp. 364-412.

[22] W. Salmon, (1971), pp. 29-87.

[23] For further study of the S-R model see Salmon, (1971); Meixner, (1979); and Lehman, (1972).

CHAPTER FIVE

[1] This is all true, I believe, without reverting to idealism or perhaps even to Kant's subjectivist epistemology. See Hanson regarding "Theory Ladenness" in *Patterns of Discovery* (1972), p. 54f. (The term 'Theory Loaded' is also used.)

[2] Barbour, p. 196.

[3] This argument is taken from Barbour, p. 196.

[4] Barbour, p. 197. Barbour uses 'lawful' instead of 'repeatable.'

[5] See Chapter One, footnote 26. The name for this form of valid deduction is *modus tollens.*

[6] Technically referred to as the Fallacy of Affirming the Consequent.

[7] I say "at most a few" because as we saw in Chapter Two (p. 51), dogmatism has an important function in science and may prevent the rejection of an old theory on the basis of only one falsifying event, however serious it may be.

[8] Of course it is simplistic to limit the alternatives this radically. There have been many attempts to resolve the problem of induction. These have included the work of Salmon in Brody and perhaps most notably, the work of Nelson Goodman esp. (1965), pp. 59-83. Goodman's approach has been to make the old problem of induction merely a pseudo-problem. It arises according to Goodman, because we insist on distinguishing the justification of induction from the description of everyday inductive practice. When we avoid this distinction the problem disappears. But Goodman acknowledges that the Ravens' Paradox discussed in the next section below requires some alteration of the confirmation theory we have been discussing, and furthermore that a new problem for confirmation arises. This "New Riddle of Induction" concerns the problem of "grue" and is well explained in Goodman, pp. 72-81. Again, this problem has received considerable attention with ambiguous success (see *e.g.*, P. Achinstein and S. Barker, in Brody, pp. 517-527 and S. Shoemaker, [1975]). So whether one says the old problem of induction is unanswered or the new riddle of induction is unanswered, in either case, confirmation seems to introduce a weakness or uncertainty into the scientific activity.

[9] Cf. B. Russell, in Brody, pp. 572-577, and note 27 for Chapter One above.

[10] This paradox, as well as the Problem of Grue (see note 8), and their at-

tempted solutions, receive clear treatment in Lambert and Brittan, pp. 76-88 in a section entitled "The Paradoxes of Confirmation." Cf. Goodman, p. 70; Hempel, (1943); and Hempel (1945), reprinted in Brody, pp. 384-409.

[11]More precisely this should be written (x) $(R_x \to B_x)$ and read "For all x's, if x is a Raven then x is Black." The material implication ($\to$) used here is defined in such a way that it is true not only when x is R and x is B but also whenever x is not R regardless of whether x is also B. This can be summarized in the truth table:

	R	B	R$\to$B
1.	T	T	T
2.	T	F	F
3.	F	T	T
4.	F	F	T

Lines 3 and 4 may seem counterintuitive, but they reflect the logical empiricist's desire (esp. Russell in *Principia Mathematica*) to capture the *minimum* meaning of implication among many possible senses it may have. This minimum is that it is not possible to affirm the antecedent and deny the consequence of an implication. For R$\to$B this means − (R $\cdot$ − B). (See I. Copi *An Introduction to Logic*. 5th ed. [N.Y.: MacMillan; 1978], pp. 277-285.)

[12]Of course Hume would remind us that constant conjunction is all we can ever know about causality, but he makes certain presuppositions of his own with which one may argue (*e.g.*, sensory atomism). This is not a place to critique Hume, so for our purpose we must at least act as though we can distinguish the two ('correlation' and 'connection') even if not exclusively.

[13]The Laws of Inertia, of Accelerations, and of Reaction.

[14]Pierre Simon Laplace, p. 4.

[15]This account and much of what follows has been based on I. Barbour's excellent Chapter 10, pp. 273-316.

[16]There is perhaps a legitimate question to be raised about whether one is justified in transferring principles used in physics onto the nature of the world; i.e. in moving from physics to metaphysics. The positivists deny this approach and point to Newtonian physics as an example of this practice gone astray.

[17]Here again the problem of moving to metaphysics arises. How do we get beyond what we know to the nature of "things-in-themselves"?

BIBLIOGRAPHY

Achinstein, P. *Concepts of Science.* Baltimore: Johns Hopkins University Press, 1968.

————. *Law and Explanation.* Oxford: Clarendon, 1971.

Achinstein, P. and Barker, S. "On the New Riddle of Induction." Brody, *Readings in the Philosophy of Science.* pp. 517-527.

Bacon, F. *The New Organon.* New York: Bobbs-Merrill, 1960.

Barker, S. F. *Induction and Hypothesis.* Ithaca: Cornell University Press, 1957.

Barbour, I. *Issues in Science and Religion.* New York: Harper Torchbook, 1966.

Blackwell, R. *Discovery in the Physical Sciences.* Notre Dame: University of Notre Dame Press, 1969.

Brody, B. (ed.). *Readings in the Philosophy of Science.* Englewood Cliffs: Prentice-Hall, 1970.

Brown, H. I. *Perception, Theory, and Commitment.* Chicago: University of Chicago Press, 1977.

Bronowski, J. *The Ascent of Man.* Boston: Little, Brown & Co., 1973.

Butterfield, H. *The Origins of Modern Science,* rev. ed. New York: MacMillan, 1965.

Campbell, N. R. *Physics and Elements.* Cambridge: Cambridge University Press, 1920.

————. *What is Science?* New York: Dover, 1952.

Collins, J. *Interpreting Modern Philosophy.* Princeton: Princeton University Press, 1972.

Dampier, W. C. *A History of Science.* Cambridge: Cambridge University Press, 1966.

de Santillana, G. *The Crime of Galileo.* Chicago: University of Chicago Press, 1955.

Doyle, A. C. *The Complete Sherlock Holmes.* 2 vols. Garden City: Doubleday, 1930.

Feyerabend, P. K. "How to Be a Good Empiricist." Brody, *Readings in the Philosophy of Science.* pp. 319-342.

————. *Against Method.* London: New Left Books, 1975.

Goodman, N. *Fact, Fiction, and Forecast.* Indianapolis: Bobbs-Merrill, 1965.

Hanson, N. R. *Patterns of Discovery.* Cambridge: Cambridge University Press, 1972.

————. "Is there a Logic of Scientific Discovery?" Brody, *Readings in the Philosophy of Science.* pp. 620-633.

131

Hardy, J. T. *Science, Technology, and the Environment.* New York: Saunders, 1975.

Hempel, C. G. *Aspects of Scientific Explanation.* New York: Free Press, 1965.

_____. "A Purely Syntactical Definition of Confirmation." *Journal of Symbolic Logic,* 8(1943): 122-143.

_____. *Philosophy of Natural Science.* Englewood Cliffs: Prentice-Hall, 1966.

_____. "Studies in the Logic of Confirmation." *Mind,* 54(1945): 1-26, 97-121.

Hempel, C. G. and Oppenheim, P. "Studies in the Logic of Explanation." *Philosophy of Science,* 15(1948): 135-175.

Hesse, M. *Models and Analogies in Science.* Notre Dame: University of Notre Dame Press, 1970.

_____. *The Structure of Scientific Inference.* Berkeley: University of California Press, 1974.

Huxley, T. H. "The Progress of Science." *Darwiniana,* I. New York: Appleton, Century, and Crofts, Inc., 1896. pp. 42-129.

Koestler, A. *The Act of Creation.* London: Hutchison, 1964.

Kuhn, T. S. *The Structure of Scientific Revolutions,* IEUS II/2, 2nd ed. Chicago: The University of Chicago Press, 1969.

_____. "The Function of Dogma in Scientific Research." Brody, *Readings in the Philosophy of Science.* pp. 356-374.

Lakatos, I. and Musgrave, A. (eds.). *Criticism & The Growth of Knowledge.* Cambridge: Cambridge University Press, 1978.

Laplace, P. S. *A Philosophical Essay on Probabilities,* 6th ed., trans. F. W. Truscott and F. L. Emory. New York: Dover, 1961.

Lambert, K. and Brittan, G., Jr. *An Introduction to the Philosophy of Science.* Englewood Cliffs: Prentice-Hall, 1970.

Laudan, L. *Progress and Its Problems.* Berkeley: University of California Press, 1977.

Lehman, H. "Statistical Explanation." *Philosophy of Science,* 39(1972): 500-506.

Levi, W. A. *Philosophy as Social Expression.* Chicago: University of Chicago Press, 1974.

McCain, G. and Segal, E. *The Game of Science.* Monterey: Brooks/Cole, 1969.

Meixner, J. "Homogeneity and Explanatory Depth." *Philosophy of Science,* 46(1979): 366-381.

Morphet, C. *Galileo and Copernican Astronomy,* SISCON. Boston: Butterworths, 1977.

Nagel, E. *The Structure of Science.* New York: Harcourt-Brace, 1961.

Newton, I. *Philosophiae Naturalis Principia Mathematica.* ed. Andre Koyre and I. B. Cohen, 2 vols., 3rd ed. Cambridge: Cambridge University Press, 1972.

_____. *Optics.* reprinted in *Great Books of the Western World,* Vol. 34. Chicago: Encyclopedia Brittanica, Inc., 1952.

Nidditch, P. H. *The Philosophy of Science.* Oxford: University Press, 1968.

Polanyi, M. *Personal Knowledge.* New York: Harper and Row, 1964.

_____. *The Tacit Dimension.* Garden City: Doubleday, 1957.

Popper, K. *The Logic of Scientific Discoveries.* New York: Harper and Row, 1959.

_____. *Conjectures and Refutations.* New York: Harper and Row, 1968.

Peirce, C. S. *Essays in the Philosophy of Science.* ed. Vincent Tomas. Indianapolis: Bobbs-Merrill, 1957.

Russell, B. "On Induction." Brody, *Readings in the Philosophy of Science.* pp. 572-577.

Salmon, W. C. "Discussion: Reply to Lehman." *Philosophy of Science,* 40 (1973): 397-402.

_____. "Inductive Inference." Brody, *Readings in the Philosophy of Science.* pp. 597-618.

_____ (ed.). *Statistical Explanation and Statistical Relevance.* Pittsburgh: Pittsburgh University Press, 1971.

Schlesinger, G. *Confirmation and Confirmability.* Oxford: Clarendon, 1974.

Shoemaker, S. "On Projecting the Unprojectible." *Philosophical Review,* 84 (1975): 178-219.

Sinclair, W. J. *Semmelweiss: His Life and His Doctrine.* Manchester: University Press, 1909.

Snow, C. P. *Two Cultures: And a Second Look.* Cambridge: Cambridge University Press, 1965.

Suppe, F. *The Structure of Scientific Theories.* Urbana: University of Illinois Press, 1977.

Taton, R. (ed.). *History of Science: The Beginnings of Modern Science 1450-1800.* New York: Basic Books, 1964.

Toulmin, S. "The Evolutionary Development of Natural Science." *American Scientist,* 55(1967): 456-471. Reprinted in Truitt *et. al.,* pp. 106-117.

_____. *Foresight and Understanding.* New York: Harper Torchbook, 1961.

_____. *The Philosophy of Science.* New York: Harper, 1960.

Truitt, W. H. *et. al. Science, Technology, and Freedom.* Boston: Houghton-Mifflin, 1974.

Wright, L. *Teleological Explanations.* Berkeley: University of California Press, 1976.

INDEX

Terms and Names:

Abduction 13, 14, 16, 17, 20, 22, 24, 25
Accidental generalizations 64-65
Analogy 17, 25, 73-75
Anomaly 51
Aristotle 8, 91, 92
    Teleological explanation (33), 72, 73
    Motion 33-36, 49, 50
    Astronomy 39-40, 42, 44-45
Bacon, Francis 14, 16, 92
Balmer, 59
Bohm, D. 108
Bohr, N. 109
Bridge principles 60-61, 62-64, 66
Childbed fever 6-8
Circularity of induction 23, 97-102
Common sense 11-12
Concepts
    Theoretical 59-64, 65-69
    As a purpose of science 10
Confirmation 12, 21-23, 59, 94, 97-102
Consensus 48
Copernicus 42-44, 44-47, 50
Correspondence rules (60)
Counterfactual conditionals 65
Covering-law model 71, 78-87
Creativity 6, 16, 17, 26, 27, 28, 38
Crises in scientific development 49, 52
Darwin, C. 52
DeBroglie, L. 108
Deduction 20, 24, 79-82, 84-85, 98
Deferent 40, 41, 42
Definition of theoretical terms 65-69
    Implicit 66-67

Explicit 66-67
Discovery 12-18, 22, 23, 24-25, 49
D-N (Deductive-Nomological) model 78-82
    Counter-examples 85-87
Dogma (of scientists) 48, 51
D-S (Deductive Statistical) (83)
Eccentric 40-41, 42
Einstein, A. 11, 51, 108-109
Elegance 43, 47, 50
Enthymemes 82
Epicycle 40, 41, 42, 43
Equant 41, 42
Error, a cause of uncertainty 96-97
Evolutionary 48-50, 51-54
Experiment 19, 20-21, 22, 28, 97, 98, 99
Explanandum 72, 73-74, 76-87
Explanans 72, 73, 74, 76-85
Explanation 12, 13, 71-88
    Analogical 74-75, 78
    D-N 78-82
    I-S 82-83
    Prediction similarity 84, 85
    S-R 87
    Teleological (33), 72-73, 78
Explanatory power 46-47, 60
Extraordinary science 51-52
Faith 48, 93, 102
Falsification 22, 23, 98
Familiarity 74-75
Formula 59-60, 61, 65
Galileo 8, 36-37, 44-46, 48, 50, 75, 92
Gaussian distribution of error 97
Generalization 11, 13, 15, 16, 57, 58
    Universal 64, 83

135

Parentheses ( ) indicate indirect references or references in footnote for that
page.

ABOUT THE AUTHOR

V. James Mannoia received his B. S. in physics from the Massachusetts Institute of Technology in 1971. He spent one year in the philosophy of science graduate program at St. Louis University, and then in 1975, completed his M. A. and Ph. D. in philosophy at Washington University—St. Louis. He taught part-time at Washington University and Greenville College, served for three years as Assistant Professor of physics at Grove City College in Pennsylvania, and is currently Assistant Professor of philosophy at Westmont College in Santa Barbara, California.

Dr. Mannoia worked in several large research groups while an undergraduate at M.I.T., including the laser optics group of Professor Ali Javan, where he did research on cadmium vapor systems. His Ph.D. dissertation in the area of process metaphysics is entitled *Whitehead's Ontological Principle: A Defense and Interpretation*. While at Washington University he received a Hans Reichenbach Fellowship in the philosophy of science and was Editorial Assistant for the *Journal of the History of Philosophy*. During the summer of 1978 he was appointed Post-Doctoral Fellow at Johns Hopkins University to do research on scientific explanation with Professor Peter Achinstein sponsored by the National Endowment for the Humanities. His continuing interests include process metaphysics, scientific explanation, and computer assisted logic curricula.